Field Columbian Museum

Publication 120

Geological Series
Vol. III, No. 5.

ANALYSES OF IRON METEORITES COMPILED AND CLASSIFIED

BY

Oliver Cummings Farrington,
Curator, Department of Geology

Chicago, U. S. A.
March 1, 1907.

ANALYSES OF IRON METEORITES
COMPILED AND CLASSIFIED

BY

OLIVER CUMMINGS FARRINGTON

Chemical analyses may be collected and grouped for purposes of record and of comparison. For the first purpose it is desirable that all known analyses of the substances under consideration be collected; for the second, only those known to be complete and reliable are needed. A combination of these two purposes may perhaps be gained, however, by collecting all analyses and leaving to the judgment of the investigator the selection of those suited for the study of any particular phase of the subject. This plan is practically that which has been adopted in presenting the analyses here collected. In many cases obviously incomplete analyses are given because they represent all that is known of the chemical constitution of the meteorite in question, or because they mark a stage in its study. On the other hand, analyses which amount to little more than a qualitative determination of the presence of iron and nickel, or whose connection with a particular meteorite is uncertain, are omitted. About three hundred and sixty analyses are here included, and it is believed that they comprise practically all of importance that have been made of iron meteorites. When more than one analysis of a meteorite is given, the analyses have been arranged chronologically. For the most part the later analyses are the most complete and reliable ones, though this is not always the case. Thus those by J. Lawrence Smith, although made thirty and in some cases forty years ago, accord well with what is known of the constitution of the iron meteorites at the present day and may be considered generally accurate and reliable. The same is true of analyses by Jackson, Berzelius, Damour, and others. As shown later, the relations between structure and composition brought out by the analyses as here grouped are so definite that at the present time a knowledge of the structure of a meteorite will give a more accurate idea of its composition than inferior chemical analyses. The general plan of arrangement which has been adopted

59

for the analyses is that now generally known as the Rose-Tschermak-Brezina classification.   This seemed the classification most desirable to employ on account of its wide use, and when it was found, as will be seen by the tables, that the chemical constitution of the meteorites follows its main divisions, its adaptation to the work in hand seems unquestionable.   Under each group of the classification the arrangement of the meteorites is alphabetical.   Synonyms of the meteorite names will be found on subsequent pages.   The characterization of the meteorite groups which head the tables have largely been summarized from Cohen.*   In considering the analyses it should be realized that some of the groups are much better known than others. Thus the ataxites and hexahedrites were thorougly studied by Cohen and their composition satisfactorily determined.   The fine octahedrites have also been mostly investigated.   The coarse and medium octahedrites, however, though more numerous than the groups just mentioned, are but imperfectly known and need detailed modern study.   In a list following the tables meteorites of which no analysis is known are marked with an asterisk.   These number about forty.   In addition, many meteorites, analyses of which are reported in the tables, have never in fact been properly studied.   The only extensive list of analyses of iron meteorites which has lately been previously compiled of which the writer is aware is that of Wadsworth, published in 1884.†   This list includes one hundred and ninety-three analyses of iron meteorites and terrestrial irons, arranged in order of the per cent of nickel.   No further attempt at classification is made. While Wadsworth's list is fairly complete as regards older analyses, it includes several pseudo-meteorites, and obviously does not adequately represent present knowledge.

The first recorded attempt at analysis of an iron meteorite is probably to be found in the examination in 1802, by Count de Bournon,‡ of some so-called native irons from Bohemia, Senegal, and South America.   In these Count de Bournon found percentages of nickel ranging from five to ten per cent, but it is stated by Howard elsewhere in the paper that owing to lack of knowledge of the peculiarities of nickel these figures are little more than estimates.   The next year Klaproth§ reported one and one-half to three and one-half per cent of nickel in the iron meteorite of Hraschina, and expressed the opinion that the presence of nickel might serve as a criterion for

---

* Meteoritenkunde, Heft III.

† The Rocks of the Cordilleras, Memoirs Museum Comparative Zoölogy, Cambridge, Mass , Vol. XI, Part I, pp. vi-xvi, Table II.

‡ Phil. Trans. Roy. Soc., London, 1802.

§ Abhandl. Akad. Wiss., Berlin, 1803, 21-41.

judging the meteoric origin of a body. Cobalt was reported by
Stromeyer in the iron meteorite of Cape of Good Hope in 1816,[*] and
copper by the same investigator in 1833.[†] Stromeyer expressed the
belief that copper was, with cobalt, a constant ingredient of meteoric
nickel-iron, and this conclusion was later corroborated by Smith[‡] on
the basis of more than one hundred analyses. Chromium was dis-
covered as a component of meteoric nickel-iron by Laugier in 1817.[§]
The presence of manganese and tin in meteoric nickel-iron was also
early reported. The presence of other metals or semi-metals reported
at different times, such as zinc, lead, arsenic, and antimony, has not
been confirmed, while the presence of aluminum, calcium, magnesium,
potassium, and sodium, noted by several analysts, is doubtless to be
referred to small quantities of silicates which either formed a constituent of the meteorite, as in Tucson, Tula, etc., or accidentally contami-
nated the material analyzed. The occurrence of phosphorus in meteoric nickel-iron seems first to have been noted by Berzelius in the
undissolved residue of Bohumilitz. It was similarily reported by
analysts who followed Berzelius, but percentages were not commonly
given until later times. Sulphur was early noted as an ingredient of
meteoric stones and later of irons. Since it occurred as a soluble
constituent, it was more often reported in the early analyses than
phosphorus. The presence of carbon as graphite was noted by Ten-
nant[¶] in 1806 in the Cape of Good Hope meteorite. Being, like the
phosphides, insoluble, its presence was often later reported in insoluble residues, but its amount was rarely given. Silicon, as reported in
the earlier analyses, whether as metal or oxide, is probably for the
most part to be referred to accessory silicates. With later methods,
however, its detection in small quantities as an ingredient of the
nickel-iron has become possible. The first detection of chlorine as
an essential constituent of iron meteorites seems to have been by Jackson in 1838,[**] in the meteorite of Limestone Creek. Its presence has
been occasionally but not commonly reported by later analysts. Determinations of specific gravity of the iron meteorites examined seem to
have been common. While these are probably for the most part fairly
reliable, some of the values reported are too anomalous to seem
trustworthy.

[*] Gottingische Gelehrte Anzeigen, 1816, 2041-2043.
[†] Gottingische Gelehrte Anzeigen, 1833, 369-370.
[‡] Am. Jour. Science, 1870 (2), 49, 332.
[§] Ann. Chem. Pharm., 1817, IV, 363-366.
Pogg. Ann., 1832, XXVII, 128-132.
[¶] Tillochs Phil. Mag., London, 1806, XXV, 182.
[**] Am. Jour. Science (1), 34, 332-337.

IRON METEORITES.

These are meteorites consisting essentially of nickel-iron. Most of them contain, in addition, an appreciable amount of sulphides, carbides, and phosphides, but the presence of silicates in quantity removes a meteorite from this class. The iron meteorite of Tucson contains about five per cent of forsterite, and the meteorites of Kodaikanal, Persimmon Creek, and Tula also contain silicate aggregates, but in small quantities. In general, it may be said that if the quantity of silicate grains exceeds five per cent the meteorite is not considered as belonging to the class of iron meteorites. About two hundred and fifty iron meteorites are now recognized, the exact number being indeterminate on account of differences of opinion as to identity of origin in several cases. The chief divisions of iron meteorites, according to the Rose-Tschermak-Brezina classification, are hexahedrites, octahedrites, and ataxites. These are sub-divided as follows:

CLASSIFICATION OF IRON METEORITES ACCORDING TO ROSE, TSCHERMAK, BREZINA, AND COHEN

  I.   Hexahedrites.
        A.   Normal hexahedrites.
        B.   Brecciated hexahedrites.
  II. Octahedrites.
        A.   Normal octahedrites.
            1. Coarsest octahedrites.
            2. Coarse octahedrites.
            3. Medium octahedrites.
            4. Fine octahedrites.
                a. Prambanan group.
                b. Rodeo group.
            5. Finest octahedrites.
                a. Salt River group.
                b. Tazewell group.
                c. Cowra and Victoria West.
        B.   Hammond octahedrites.
        C.   Brecciated octahedrites.
  III. Ataxites.
        A.   Nickel-poor ataxites.
            1. Siratik group.
            2. Nedagolla group.
            3. Rafruti group.

B.   Nickel-rich ataxites.
    1.   Smithland group.
    2.   Cristobal group.
    3.   Octibbeha.
C.   Ataxites with forsterite.
D.   Ataxites with cubic streaks.

The iron meteorites enumerated according to groups sum up as follows:

Octahedrites:

Coarsest	13
Coarse	30
Medium	98
Fine	33
Finest	14
Brecciated	6
Hammond	3
Unclassified	4
	201
Ataxites	30
Hexahedrites	17
Total	248

ALPHABETICAL LIST OF IRON METEORITES.

The following is an alphabetical list of iron meteorites, showing the classification of each.   An asterisk indicates that no analysis of the meteorite is reported.

Abert Iron.........Medium octahedrite
*Adargas...........Medium octahedrite
Algoma ....Medium octahedrite
Alt Biela..........Fine octahedrite
*Amates ....Medium octahedrite
Angara...........Medium octahedrite
*Apoala ....Fine octahedrite
Arispe ....Coarsest octahedrite
Arlington ....Medium octahedrite
Asheville..........Medium octahedrite
Auburn...........Hexahedrite
Augustinowka......Fine octahedrite

Babb's Mill........Ataxite
Bacubirito ....Finest octahedrite

Bald Eagle........Medium octahedrite
Ballinoo...........Finest octahedrite
Barranca Blanca...Brecciated octahedrite
Beaconsfield.......Coarse octahedrite
Bear Creek........Fine octahedrite
Bella Roca.........Fine octahedrite
Bendego...........Coarse octahedrite
Bethany...........Fine octahedrite
Billings ....Coarse octahedrite
Bingera ....Hexahedrite
Bischtube.........Coarse octahedrite
Black Mountain....Coarse octahedrite
*Blue Tier.........Medium octahedrite
Bohumilitz ....Coarse octahedrite

Boogaldi...........Fine octahedrite
Botetourt..........Ataxite
Braunau...........Hexahedrite
Bridgewater ......Fine octahedrite
Buckeberg ....Fine octahedrite
Burlington ....Medium octahedrite
Butler.............Finest octahedrite

Cabin Creek......Medium octahedrite
Cacaria ....Hammond octahe-
                   drite
Cachiyuyal.........Medium octahedrite
Cambria...........Fine octahedrite
Campo del Cielo...Ataxite
Canton ....Coarsest octahedrite
Canyon Diablo.....Coarsest octahedrite
Canyon City......Coarse octahedrite
Cape of Good Hope.Ataxite
Caperr ....Medium octahedrite
Cape York........Medium octahedrite
Carlton ....Finest octahedrite
Carthage ....Medium octahedrite
Casas Grandes.....Medium octahedrite
*Casey County......Coarsest octahedrite
Central Missouri...Coarsest octahedrite
*Chañaral ....Coarse octahedrite
*Charcas ....Medium octahedrite
*Chambord.........
Charlotte.........Fine octahedrite
Chesterville........Ataxite
*Chichimeguilas.....
Chilkoot...........Medium octahedrite
Chulafinnee.......Medium octahedrite
Chupaderos ....Fine octahedrite
Cincinnati ....Ataxite
Cleveland.........Medium octahedrite
Coahuila ....Hexahedrite
Colfax............Medium octahedrite
Coopertown.......Medium octahedrite
Cosby Creek......Coarse octahedrite
Costilla...........Medium octahedrite
Cowra ....Finest octahedrite
*Cranberry Plains...Octahedrite
Cranbourne.......Coarse octahedrite
Cuba.............Medium octahedrite
Cuernavaca.......Fine octahedrite

Dalton ....Medium octahedrite
Deep Springs .....Ataxite.............
Dehesa...........Ataxite

*Dellys ....Medium octahedrite
Denton County.....Medium octahedrite
Descubridora ......Medium octahedrite
De Sotoville ......Hexahedrite
Duell Hill.........Coarse octahedrite

Elbogen ....Medium octahedrite
El Capitan... .....Medium octahedrite
*El Tule ....Medium octahedrite
*Emmitsburg ,......Medium octahedrite

Forsyth County....Ataxite
Fort Duncan.. ....Hexahedrite
Fort Pierre ....Medium octahedrite
Franceville.......Medium octahedrite
Frankfort.........Medium octahedrite

Glorieta ....Medium octahedrite
Grand Rapids......Fine octahedrite
Greenbrier County..Coarse octahedrite
Groslee...........Fine octahedrite
Guilford County....Medium octahedrite

Hammond ....Hammond octahe-
                 drite
*Haniel el-Beguel...Medium octahedrite
Hassi Jekna........Fine octahedrite
*Hayden Creek.....Medium octahedrite
Hex River ....Hexahedrite
Holland's Store....Hexahedrite
Hopewell Mounds..Medium octahedrite
Hopper ....Medium octahedrite
Hraschina.........Medium octahedrite

*Ilimae.............Medium octahedrite
Illinois Gulch......Ataxite
Indian Valley......Hexahedrite
Iquique............Ataxite.............
Iredell.............Hexahedrite
Ivanpah...........Medium octahedrite

*Jackson County....Medium octahedrite
Jamestown ....Fine octahedrite
Jennie's Creek .....Coarse octahedrite
Jewel Hill.........Fine octahedrite
Joel's Iron.........Medium octahedrite
Joe Wright.......Medium octahedrite
Jonesboro.........Fine octahedrite
Juncal ....Medium octahedrite

Kendall County....Hexahedrite
Kenton County.....Mediumoctahedrite

*Kodaikanal........Fine octahedrite
Kokomo...........Ataxite
Kokstad..........Medium octahedrite

La Caille.........Medium octahedrite
Lagrange.........Fine octahedrite
Laurens County....Finest octahedrite
Lenarto...........Medium octahedrite
Lexington County..Coarse octahedrite
Lick Creek.......Hexahedrite
Limestone Creek...Ataxite
Linville...........Ataxite
Locust Grove......Ataxite
*Lonaconing.......Coarse octahedrite
Losttown..........Medium octahedrite
*Lucky Hill........Medium octahedrite
Luis Lopez.......Medium octahedrite

*Madoc...........Fine octahedrite
Magura...........Coarse octahedrite
Mantos Blancos....Finest octahedrite
Marshall County...Medium octahedrite
Mart.............Finest octahedrite
Matatiela.........Medium octahedrite
Mazapil..........Medium octahedrite
Merceditas........Medium octahedrite
Misteca..........Medium octahedrite
*Moctezuma.......Medium octahedrite
*Mooranoppin......Coarsest octahedrite
Moonbi...........Fine octahedrite
Morito...........Medium octahedrite
Morradal.........Ataxite
Mount Joy........Hexahedrite
*Mount Stirling....Coarse octahedrite
Mungindi.........Finest octahedrite
Murfreesboro......Medium octahedrite
Murphy...........Hexahedtite

*Nagy-Vazsony.....Medium octahedrite
Narraburra Creek..Finest octahedrite
Nedagolla.........Ataxite
Nejed............Medium octahedrite
Nelson County.....Coarsest octahedrite
Nenntmannsdorf..Ataxite
N'Goureyma.......Brecciated octahedrite
Niagara...........Coarse octahedrite
*Nochtuisk.........Coarse octahedrite
*Nocoleche.........Medium octahedrite

Oktibbeha County..Ataxite

Orange River......Medium octahedrite
*Oroville..........Medium octahedrite
Oscuro Mountains..Coarse octahedrite

Pan de Azucar.....Coarse octahedrite
*Persimmon Creek..Brecciated octahedrite
Petropawlowsk....Medium octahedrite
Pittsburg..........Coarsest octahedrite
Plymouth.........Medium octahedrite
Ponca Creek......Coarsest octahedrite
Prambanan........Fine octahedrite
Primitiva.........Ataxite
Puquois..........Medium octahedrite
Putnam County....Fine octahedrite

Quesa............Fine octahedrite

Rafruti..........Ataxite
*Rancho de la Pila..Medium octahedrite
Rasgata..........Ataxite
Red River........Medium octahedrite
Reed City........Hammond octahedrite
Rhine Valley.....Medium octahedrite
Rodeo............Fine octahedrite
Roebourne........Medium octahedrite
*Rosario...........Octahedrite
Rowton...........Medium octahedrite
Ruff's Mountain....Medium octahedrite
Russel Gulch......Fine octahedrite

Sacramento Mountains...........Medium octahedrite
St. Francois County.Coarse octahedrite
St. Genevieve County..............Fine octahedrite
Salt River........Finest octahedrite
San Angelo........Medium octahedrite
San Cristobal......Ataxite
San Francisco del Mezquital........Ataxite
*Santa Apolonia....
Santa Rosa........Brecciated octahedrite
Sao Juliao.........Coarsest octahedrite
Sarepta...........Coarse octahedrite
Schwetz..........Medium octahedrite
Scottsville........Hexahedrite
Seelasgen..........Coarsest octahedrite
Seneca Falls......Medium octahedrite

Shingle Springs....Ataxite
*Sierra Blanca......Coarse octahedrite
Silver Crown ......Coarse octahedrite
Siratik............Ataxite
Smithland........Ataxite
Smith's Mountain..Fine octahedrite
Smithville ....Coarse octahedrite
Ssyromolotow......Medium octahedrite
Staunton ....Medium octahedrite
Summit...........Hexahedrite
Surprise Springs...Medium octahedrite

Tabarz...........Coarse octahedrite
*Tajgha...........Medium octahedrite
*Tanogami........Medium octahedrite
Tazewell..........Finest octahedrite
*Teocaltiche ......Octahedrite
Ternera..........Ataxite
Thunda ....Medium octahedrite
Thurlow..........Fine octahedrite
*Tlacotepec ......Octahedrite
Toluca ....Medium octahedrite
Tonganoxie ....Medium octahedrite
Toubil ....Medium octahedrite
Trenton..........Medium octahedrite
Tucson...........Ataxite

Tula ....Brecciated octahe-
    drite
*Union County......Coarsest octahedrite
Ute Pass ....Coarsest octahedrite

Varas ....Fine octahedrite
Victoria ....Medium octahedrite
Victoria West......Finest octahedrite
*Wallen's Ridge....Coarse octahedrite
Walker County.....Hexahedrite
Weaver ....Ataxite
Welland..........Medium octahedrite
*Werchne Dniep-
    rowsk ....Finest octahedrite
Werchne Udinsk ..Medium octahedrite
Wichita County....Coarse octahedrite
Willamette ....Medium octahedrite
Wooster..........Medium octahedrite

Yanhuitlan........Fine octahedrite
Yardea Station.....Medium octahedrite
*York.............Medium octahedrite
Youndegin ....Coarse octahedrite

Zacatecas.........Brecciated octahe-
    drite

SYNONYMS.

The following are synonyms of the iron meteorites given in the preceding list:

Aeriotopos...........Bear Creek
Agram..............Hraschina
Ainsa ....Tucson
Albuquerque ....Glorieta
Allen County........Scottsville
Amakaken...........Caperr
Arva...............Magura
Atacama, 1858.......Joel's Iron
Atacama, 1874.......Cachiyuyal
Augusta County......Staunton

Bahia ....Bendego
Baird's Farm........Asheville
Bates County........Butler
Batesville...........Joe Wright
Bonanza ....Coahuila
Brazos River.........Wichita
Butcher Iron........Coahuila

Caille ....La Caille
Caney Fork..........Carthage
Carleton Iron........Tucson
Catorze ....Descubridora
Chatooga County......Holland's Store
Cherokee County,1867.Losttown
Cherokee County,1894.Canton
Chilkat.............Chilkoot
Claiborne...........Lime Creek
Cocke County.......Cosby Creek
Concepcion..........Adargas
Cross Timbers.......Red River
Crow Creek.........Silver Crown

Dakota ....Ponca Creek

Ellenboro ....Colfax

Floyd County.......Indian Valley

Floyd Mountain......Indian Valley

Great Fish River.....Bethany
Green County........Babb's Mill

Hamilton County.....Carlton
Hastings County......Madoc
Hauptmannsdorf .....Braunau
Henry County, 1857...Locust Grove
Henry County, 1889...Hopper
Honduras...........Rosario
Howard County......Kokomo

Independence County.Joe Wright
Independence.......Kenton County
Iron Creek..........Victoria

Johnson County......Cabin Creek

Knoxville...........Tazewell

La Primitiva.........Primitiva
Lea Iron............Cleveland
Lime Creek, 1832.....Walker County
Lime Creek, 1834.....Limestone Creek
Lion River..........Bethany
Lockport............Cambria

Miller's Run ....Pittsburg
Muchachos..........Tucson

Mukerop............Bethany

Netschaevo..........Tula

Obernkirchen........Buckeberg
Oldham County.......La Grange

Penkarring Rock.....Youndegin

Ranchito............Bacuburito

Saltillo.............Coahuila
Sanchez Estate.......Coahuila
San Gregorio ....Morito
Saskatchewan........Victoria
Senegal.............Siratik
Serrania de Varas.....Varas
Sierra de la Ternera..Ternera
Southeast Missouri....St. Francois County

Teposcolula.........Yanhuitlan
Tocavita............Santa Rosa
Tombigbee River.....De Sotoville
Tucuman ....Campo del Cielo

Waldron's Ridge.....Wallen's Ridge
White Sulphur
  Springs...........Greenbrier County
Whitfield County.....Dalton
Wohler's Iron.......Campo del Cielo

ANALYSES OF IRON METEORITES.

I. HEXAHEDRITES.

The hexahedrites are characterized by cubic cleavage and Neumann lines.   They consist of the single alloy kamacite, the composition of which, Fe_{14} Ni, shows a close approximation to the iron-nickel content of the hexadedrites.   The content of phosphorus in the hexahedrites is usually relatively high, ¼ to ½ per cent.   This appears char-

NORMAL

Name.	Fe.	Ni.	Co.	Cu.	Cr.	P.	S.	C.	Si.	Cl.	In-sol.	Miscellaneous.
Auburn...........	94.58	3.01	.....	.....	.....	.13	......	.52	.....	.....	.....	....
"	94.49	4.67	1.03	.101	.024	.46	.002	.....	.....	.....	.....	....
Braunau..........	91.88	5.52	.53	.....	.....	......	......	.....	.....	.....	.....	2.07
"	93.62	5.21	.92	.07	.05	.24	.08	.09	.....	.....	.....	....
Coahuila..........	94.82	5.62	.60	.....	tr.	.29	......	.....	.....	.....	.06	
(Bonanza)...........	97.90	2.10	tr.	.....	tr.	tr.	......	.....	.....	.....	.....	Mg. tr.
(Butcher)............	92.95	6.62	.48	tr.	.....	.02						
(Saltillo)............	94.62	4.79	.60	.04	tr.	.18	tr.	.....	.....	.....	.....	
(Santa Rosa).........	96.07	3.26	.55	.....	.....	1.05						
" "	91.86	7.42	.50	.....	.....	.27						
De Sotoville.......	95.02	4.11	.40	.....	.....	.32	tr.	.16				
"	95.14	4.82		.05	.01	.29	.06	.....				
"	95.18	4.32	.69	.04	.....	.20	......	.07				
"	95.41	4.04	.74	.04	.....	.14	.05	.....	.....	.....	.02	....
Fort Duncan......	94.90	4.87		.....	.....	.23	tr.	tr.	.....	.....	.....	
" "	92.02	6.10		.....	.....	......					1.80	
" "	91.90	7.03		.....	.....	......						
" "	92.58	6.66	.73	.....	.....	.28	......	.....	.....	.....	.01	....
" " ...	94.65	4.82	1.07	.04	.04	.23	.32	.....	.....	.....	.....	....
(Sancha)............	96.04	3.11	.42	.....	.....	......	......	.....	.....	.....	.57	....
"	95.82	3.18	.35	tr.	.....	.24	......	.....	.....	.....	.....	....

acteristically in the hexahedrites in the form of rhabdite, and often constitutes 1½ to 3 per cent of their mass. Another characteristic mineral of the hexahedrites is daubreelite. Graphite and troilite are rare, although the latter mineral occurs in some members of the group in visible nodules. The hexahedrites may be divided into normal and brecciated hexahedrites, according to whether they are one or several individuals.

A. NORMAL HEXAHEDRITES.

In these hexahedrites the cleavage planes and Neumann lines run without change of direction throughout the mass.

HEXAHEDRITES.

Loss	Undet.	Total.	Sp. Gr.	Analyst.	Reference.
.....	.....	98.24	7.–7.17	C. U. Shepard	1869, A. J. S. (2), XLVII, 230-233
....	......	100.77	....	O. Hildebrand......	1905, Meteoritenkunde, III, 217
.....	......	100.00	7.782	Dutlos & Fischer....	1847, Ann. Phy. Chem., LXXII, 475-480
.....	.02	100.30	7.8516	R. Knauer..........	1905, Meteoritenkunde, III, 207
.....	.....	101.39	7.8678	E. Cohen..........	1894, Meteoreisen-Studien A. N. H., IX, 104
.....	.....	100.00	7.825	C. U. Shepard	1867, A. J. S. (2), XLIII, 385
.....	.....	100.07	7.692	J. L. Smith	1869, A. J. S. (2), XLVII, 385
.....	.....	100.22	....	O. Bürger..........	1905, Meteoritenkunde, III, 194
.....	.....	100.935	....	H. Wichelhaus......	1863, Ann. Phy. Chem., CXVIII, 631-634
.....	.....	100.05	....	N. F. Lupton.......	1885, A. J. S. (3), XXIX, 233
.....	.....	100.01	....	J. E. Whitfield......	1899, A. J. S. (4), VIII, 154
.....	.....	100.37	....	R. Knauer..........	1905, Meteoritenkunde, III, 213
.....	.....	100.50	....	Hildebrand & Cohen	Same
.....	.....	100.46	....	Knauer & Cohen....	Same
.....	.....	100.00	7.522	J. B. Mackintosh....	1886, A. J. S. (3), XXXII, 306
.....	.....	99.92	7.699	Meunier............	1887, C. R., CIV, 872-873
.....	.....	98.93	7.72	"	1893, B. S. H. N., VI, 17
.....	.....	100.26	....	E. Cohen..........	1889, Neues Jahrb., 227
.....	.02	101.19	7.84	O. Hildebrand.....	1905, Meteoritenkunde, III, 194
.....	.....	100.14	8.13	F. A. Genth........	1854, A. J. S. (2), XVII, 239-240
.....	.....	99.59	7.81	J. L. Smith..........	1855, A. J. S. (2), XIX, 160-161

Name.	Fe.	Ni.	Co.	Cu.	Cr.	P.	S.	C.	Si.	Cl.	In-sol.	Miscellaneous.
Hex River........	93.33	5.58	.84	....	....	......	......	....	....	....	....	....
" "	93.59	5.68	.66	.04	.02	.23	.08	....	....	......	.03	....
Iredell........	93.75	5.51	.52	....	....	.20	.06	....	....	....	....	....
Lick Creek........	93.00	5.74	.52	tr.	....	.36	tr.	....	....	tr.	....	....
Murphy..........	93.93	5.52	.61	.02	....	.34	......	.04	......	.06	....	....
Scottsville	94.32	5.01	tr.	....	....	.16	.34	.12	....	....	....	....
"	93.14	5.73	.99	.10	....	.15	......	....	....	....	.02	....
"	94.03	5.33	.95	.04	.02	.23	.07	....	....	......	.01	....
Walker County....	94.14	5.30	.64	.06	.05	.28	.19	....	....	....	....	....

B. BRECCIATED HEXAHEDRITES.

These hexahedrites are characterized by a structure which gives them the appearance of being aggregates of individual grains. Not only do apparent outlines of grains occur, but the directions of the Neumann lines are different on the different grains. The size of the grains differs in different falls, but is fairly uniform for meteorites of the same fall. The contour of the grains may be rounded, polygonal, elongated, or ragged, and as a rule the grains are sharply separated from one another. When the divisions between grains widen to a cleft, some accessory constituent usually occupies the gap. Accessory minerals are not, however, abundant. The presence of dau-

BRECCIATED

Name.	Fe.	Ni.	Co.	Cu.	Cr.	P.	S.	C.	Si.	Cl.	In-sol.	Miscellaneous.
Bingera...........	93.76	4.39	.57	....	....	.23	......	.14	....	......	.54	Na. tr.
"	93.50	5.54	.51	.01	....	.26	......	.03	.01	....	....	Sn. .02 Mn. Pt.Ir.tr.
Holland's Store....	93.06	5.35	1.00	....	.23	.31	.08	....	....	....	.08	....
" ...	94.60	4.97	.21	....	....	.21	tr.	tr.	....	....	....	....
Indian Valley......	93.59	5.56	.53	tr.	....	.27	.01	....	tr.	....	....	....
Kendall County...	92.65	5.64	.78	.03	.01	.34	.03	1.62	......	.01	....	....
Mount Joy	93.80	4.81	.51	.005	....	.19	.01	....	....	....	....	....
Summit..........	93.39	5.62	.58	....	....	.31	......	....	....	....	....	....

Loss.	Undet.	Total.	Sp. Gr.	Analyst.	Reference.
.....	.94	100.69	....	Cohen & Weinschenk	1891, Meteoreisen-Studien A. N. H., VI, 143
.....	.....	100.43	7.8225	R. Knauer..........	1905, Meteoritenkunde, III, 225
.....	.....	100.04	....	J. E. Whitfield......	1899, A. J. S. (4), VIII, 415-416
.....	.....	99.62	....	Smith & Mackintosh	1880, A. J. S. (3), XX, 324-326
.....	.....	100.52	7.7642	J. Fahrenhorst	1900, Meteoreisen-Studien A.N.H., XV, 368
.....	.....	99.95	7.848	J. E. Whitfield......	1887, A. J. S. (3), XXXIII, 500
.....	.....	100.13	....	Fischer............	1889, Neues Jahrb., I, 227
.....	.....	100.68	7.7959	R. Knauer..........	1905, Meteoritenkunde, III, 220
.....	.....	100.66	7.7806	O. Hildebrand......	1905, Meteoritenkunde, III, 173

breelite has not been noted, and schreibersite is not common, either in nodules or as rhabdite. The view that the brecciated hexahedrites are aggregates is not accepted by Brezina, except in the case of Kendall County. He regards the structure and cleavage of the other members of the division as uniform, and explains the varying orientation as caused by twinning. Mount Joy, placed by Berwerth, Cohen, and Brezina among the coarsest octahedrites, because of an apparent octahedral structure observed by Berwerth, seems to the present writer to belong more properly to the hexahedrites. In composition and structure it agrees fully with the hexahedrites, and it shows no trace of cohenite, a characteristic mineral of the coarse octahedrites. Its individual grains are the largest of any of the following group:

HEXAHEDRITES.

Loss.	Undet.	Total.	Sp. Gr.	Analyst.	Reference.
.....	.....	99.63	7.834–7.849	A. Liversidge.......	1882, Proc. Roy. Soc. N. S. W., XVI, 31-34
.....	.....	99.88	7.761	J. C. H. Mingaye.....	1904, Rec. Geol. Sur. N. S. W., VII, 308-310
.....	.....	100.11	....	Zaubitzer	1905, Meteoritenkunde, III, 240
.....	.....	99.99	7.801	J. E. Whitfield......	1887, A. J. S.(3), XXXIV, 472
.....	.....	99.96	7.95	L. G. Eakins........	1892, A. J. S. (3), XLIII, 424
.....	.....	101.11	....	Scherer	1900, Meteoreisen-Studien, A. N. H., XV, 387
.....	.....	99.33	....	L. G. Eakins........	1892, A. J. S. (3), XLIV, 416
.....	.....	99.90	6.949	F. P. Venable.......	1890, A. J S. (3), XL, 322

II. OCTAHEDRITES.

The meteorites of this class are the most abundant among iron meteorites.   According to the width of the lamellæ as seen in etched sections, they are divided as follows: Coarsest octahedrites, lamellæ, many mm. to 2.5 mm. in width; coarse octahedrites, lamellæ 2–1.5 mm. in width; medium octahedrites, lamellæ 1.0–0.5 mm. in width; fine octahedrites, lamellæ 0.4–0.2 mm. in width; finest octahedrites, lamellæ from 0.2 mm. down.   While no sharp line of separation can be drawn between these groups, the members of each group present as a rule characters more or less peculiar to themselves.   As compared with the hexahedrites, the octahedrites differ in structure in being made up of lamellæ arranged in accordance with the planes of the octahedron.   These lamellæ in turn are composed of two or more alloys of nickel-iron.   In composition a higher percentage of nickel-cobalt may be noted among the octahedrites, as compared with the hexahedrites, and schreibersite and troilite are far more abundant than in the hexahedrites.   Cohenite, which is not known to occur in the hexahedrites, is characteristic of certain groups of the octahe-

COARSEST

Name.	Fe.	Ni.	Co.	Cu.	Cr.	P.	S.	C.	Si.	Cl.	In-sol.	Miscellaneous.
Arispe	92.27	7.04	.....	....	....	......	.....	.....	.....	.....	....	....
Canyon Diablo	95.370	3.945	.....	....	....	.144	tr.	.....	tr.	.....	.26	....
" "	91.396	7.94	.....	....	....	.179	.004	.417	.047	.....	....	....
Canton	91.96	6.70	.50	.03	....	.11	.01	.....	tr.			
Central Missouri	94.73	4.62	.18	....	....	.44	.02	.01	.....	....	....	....
Nelson County	93.10	6.11	.41	tr.	....	.05	......	....				
Pittsburg	92.81	4.66	.39	.03	.....	.25	.04	.....	....		....	Mn....... .14
"	93.38	5.89	1.24	.05	.02	.15	.07	.....	....		....	Chromite. .07
Ponca Creek	91.74	6.53	.....	....	....	.01	......	....	....		....	Sn....... .06
" "	91.74	7.08	.....	....	....	.01	......	....	....		....	Sn....... .06
São Julião	89.39	8.27		tr.	....	.26	......	....	....		....	....
Seeläsgen	90.00	5.31	.43	.10	....	......	......	....	1.16	.....	.83	Mn...... .91
"	92.33	6.23	.67	....	....	......	......	.52	.02	.....	.18	Cu.+Sn.. .05

drites, while graphite and diamond are also largely confined to the octahedrites. Daubreelite and chromite, which are common constituents of the hexahedrites, are rare in the octahedrites. The nickel-cobalt content of the octahedrites varies from $5\frac{1}{2}$ to $15\frac{1}{2}$ per cent.

A. NORMAL OCTAHEDRITES.

In the normal octahedrites the lamellar structure extends without change of direction, except for occasional curving, through the individual. This is true even for large masses like those of Charcas, Chupaderos, and Willamette.

1. COARSEST OCTAHEDRITES.

Width of lamellæ from many millimeters down to 2.5 mm. The nickel-cobalt content is as a rule slightly higher than in the hexahedrites, reaching in some cases 7 per cent. The presence of cohenite and graphite is characteristic of the group. Canyon Diablo contains diamond. The octahedral structure and presence of lamellæ is often difficult to discern, so that some members of the group have been classed as hexahedrites.

Octahedrites.

Loss.	Undet.	Total.	Sp. Gr.	Analyst.	Reference.
.....	......	99.31	7.853	J. E. Whitfield......	1902, Proc. Roch. Acad. Sci., IV, 85
.....	......	99.719	7.703	H. Moissan.........	1904, Comptes Rendus, CXXXIX, 776
.....	......	99.983	....	Booth, Garrett & Blair	1905, Proc. Phil. Aca. Sci., LVII, 875
.....	......	99.31	....	H. N. Stokes........	1895, A. J. S. (3), L, 252-4
.....	......	100.00	....	Mariner & Hoskins..	1900, A. J. S. (4), IX, 286
.....	......	99.67	....	J. L. Smith..........	1860, A. J. S. (2), XXX, 240
.....	......	98.32	7.74	F. A. Genth.........	1876, A. J. S. (3), XII, 72-73
.....	......	100.87	....	O. Hildebrand	1903, Mitt. f. Neu Vorp. u. Rügen, XXXV,4
.....	......	98.34	7.952	C. T. Jackson.......	1863, A. J. S. (2), XXXVI, 261
.....	......	98.89	7.952	" "	1863, A. J. S. (2), XXXVI, 261
.....	......	97.92	7.783	C. v. Bonhorst	1888, Neues Jahrb., 372
.....	......	98.74	7.63 –7.71	A. Duflos..........	1848, Ann. Phy. Chem., LXXIV, 61-65
.....	......	100.00	7.73	C. Rammelsberg....	1848, Ann. Phy. Chem., LXXIV, 443-448

2. COARSE OCTAHEDRITES.

Width of lamellæ 2.0–1.5 mm.   The lamellar or octahedral struc-

COARSE

Name.	Fe.	Ni.	Co.	Cu.	Cr.	P.	S.	C.	Si.	Cl.	In-sol.	Miscellaneous.
Beaconsfield.......	92.56	7.34	.48	.02	....	.26	.04	.05	.....	.01	....	....
Bendego..........	91.90	5.71	.....	....	....	....	....	....	....	....	.46	....
"	88.46	8.59		....	....	....	....	.07	....	....	....	P. Fe. Ni. .37
Billings...........	91.99	7.38	.42	.01	....	.15	.06	.....	.08	.....	....	....
Bischtübe........	93.39	6.48	.87	.03	tr.	.05	......	....	....	.01	......	......
Black Mountain...	96.04	2.52	.....	....	....	....	.....	....	....	....	1.44	....
Bohumilitz........	94.06	4.01	.....	....	....	.81	......	....	....	....	....	C. etc.... 1.12
" .....⅔	93.12	4.74	.23	....	....	....	......	....	....	....	1.91	....
"	94.77	3.81	.20	....	....	....	......	....	....	....	2.20	....
Canyon City.......	88.81	7.28	.17	....	....	.12	......	..	....	....	....	....
" "	91.25	7.85	.17	....	....	.10	......	....	....	....	....	....
Cosby Creek......	87.00	12.00	.....	....	....	....	.50	.....	....	....	....	....
" " ....⅔	93.91	4.55	.....	....	....	....	......	....	....	....	.10	....
" "	91.64	5.85	.81	*.22	....	.19	......	....	.08	....	....	Mn.09 Graphite .80
" "	91.90	6.70	.33	....	....	.09	......	....	.18	....	....	....
" "	92.75	6 91	.51	.02	....	.37	......	....	....	....	....	....
Duell Hill........	94.24	5.17	.37	tr.	....	.14	......	....	....	....	.15	....
Greenbrier County.	91.59	7.11	.60	tr.	....	.08	......	....	....	....	....	....
Jennie's Creek	91.56	+8.31		....	....	.13	......	....	....	....	....	....
Lexington County..	92.42	6.08	.93	....	....	tr.	......	....	....	....	.26	Sn. tr.........
Magura........⅔	93.62	5.68	.....	....	....	......	......	....	....	....	....	....
"	89.42	8.61	.....	....	....	......	......	....	....	....	....	C. Cu. Si. Sch. 1.41
" .....⅔	90.91	7.32	.....	....	....	......	......	....	....	....	....	Co. C. Si., etc.,1.17
"	92.55	7.08	.51	.02	....	.24	.02	.03	......	.01	....	....
Niagara...........	92.67	7.37	.13	....	....	......	......	....	....	....	....	....
Oscuro Mountains..	90.79	7.66	.57	....	....	.27	......	.07	....	....	....	....

*Cu. Sn.   +By diff.

ture is more obvious than in the coarsest octahedrites, and the nickel-cobalt content in some members slightly higher. Cohenite and graphite are characteristic and common ingredients.

OCTAHEDRITES.

Loss.	Undet.	Total.	Sp. Gr.	Analyst.	Reference.
....	.....	100.76	....	O. Sjöström.........	1897, Sitzber. Berl. Akad., 1047
1.93	.....	100.00	7.73	Flickentscher.......	1863, Buchner, Meteorites, 144
1.96	.....	99.45	7.47	Wohler & Martius..	1867, Phipson, Meteorites, 94
....	.....	100.09	....	H. W. Nichols......	1905, A. J. S. (4), XIX, 242
....	.....	100.82	....	Scherer & Sjöstrom.	1897, Meteoreisen-Studien, V,A.N.H., XII, 55
....	.....	100.00	7.261	C. U. Shepard......	1847, A. J. S. (2), IV, 81–83
....	....	100.00	7.15	J. Steinman........	1830, A. J. S. (1), XIX, 384–386
....	.....	100.00	....	J. J. Berzelius.......	1833, Ann. Phy. Chem., XXVII, 118–132
....	.....	100.98	....	"	1853, A. J. S. (2), XV, 12
....	.....	96.38	7.1	C. U. Shepard......	1885, A. J. S. (3), XXIX, 469
.. ..	.....	99.37	7.68	J. M. Davison.......	1904, A J. S. (4), XVII, 383
.50	.....	100.00	....	G. Troost..........	1840, A. J. S. (1), XXXVIII, 254
....	.....	98.56	6.22	C. U. Shepard.......	1842, A. J. S. (1), XLIII, 354–357
....	.....	99.68	....	C. A. Joy..........	1853, Ann. Chem. Pharm., LXXXVI, 39–43
....	....	99.20	7.26	C. Bergmann.......	1857, Ann. Phy. Chem., C, 254–255
....	.....	100.56	....	J. Fahrenhorst	1900, Meteoreisen-Studien, XI, A.N.H.XV,373
....	.....	100.07	7.46	B. S. Burton........	1876, A. J. S. (3), XII, 439
.....	.12	99.50	....	L. Fletcher.........	1887, Min. Mag., VII, 183
....	.....	100.00	7.344	J. B. Mackintosh....	1886, A. J. S. (3), XXXI, 147
....	....	99.69	7.00–7.405	C. U. Shepard, Jr...	1881, A. J. S. (3), XXI, 119
....	.....	99.30	7.814	A. Patera..........	1847, Östr. Blätt. f. Lit., No. 169,–670
....	.....	99.44	7.814	"	Same
....	.....	99.30	7.01–7.22	A. Löwe...........	1849, Neues Jahrb., 199
....	.....	100.46	....	J. Fahrenhorst......	1900, Meteoreisen-Studien, XI, A.N.H.XV, 378
....	.....	100.07	7.12	J. M. Davison.......	1902, Jour. Geol., X, 518–519
....	.....	99.36	....	R. C. Hills........	1897, Proc. Colorado Sci. Soc.

Name.	Fe.	Ni.	Co.	C u	Cr.	P.	S.	C.	Si.	Cl.	In-sol.	Miscellaneous.
St. Francois County	92.10	2.60	tr.	....	....	tr.	......	tr.	tr.	.....	....	Schreibersite ..5.0
" " "	92.68	6.97	.52	.02	....	.34	.01	.....	.....	.03	.01	....
Sarepta..........	95.94	2.66	....	....	....	......	......	.....	.02	....	....	Sn.02 P.Fe.Ni. 1.32
Silver Crown......	91.57	8.31	tr.	....	....	.07	......	tr.	.....	....	....	....
Smithville.........	91.57	7.02	.62	tr.	....	.18	......	.....	.....	....	....	Res.MainlyCarb..15
Tabarz...........	92.76	5.69	.79	....	....	.86	......	.....	.....	....	....	P. Fe. Ni. .28
Wichita..........	89.99	10.01	tr.	....	....	......	......	.....	.....	....	....	....
Willamette........	91.46	8.30	....	....	....	......	......	.....	.....	....	....	....
"	91.65	7.88	.21	....	....	.09	......	.....	.....	....	....	....
Youndegin........	92.67	6.46	.55	tr.	....	.24	......	.....	.....	....	.04	Mg....... .42

3. MEDIUM OCTAHEDRITES.

Width of lamellæ 1.0-0.5 mm. More than one-third of the iron meteorites belong to this class. They present, as a rule, quite uniform characters. The lamellar structure is, as a rule, well-defined, and

MEDIUM

Name.	Fe.	Ni.	Co.	Cu.	Cr.	P.	S.	C.	Si.	Cl.	In-sol.	Miscellaneous.
Abert Iron........	92.92	6.07	.54	....	....	......	......	.....	.....	....	....	Schreibersite. .56
"	92.04	7.00	.68	....	....	.08	.01	.02	.....	....	....	Graphite. .03
Algoma	88.62	10.63	.84	....	....	.15	tr.	.....	.02	....	....	....
Angara..........	92.64	7.10	....	....	....	.16	......	tr.	.04	....	tr.	Ca. tr. Mg. .06
Arlington........	90.78	8.60	1.02	tr.	tr.	.05	......	tr.	.....	....	....	....
Asheville	96.50	2.60	....	....	....	......	......	.....	.50	.20	....	....
Bald Eagle	91.36	7.56	.70	....	....	.09	.06	.....	tr.	....	....	....
Burlington........	92.29	8.14	....	....	....	......	......	.....	.....	....	....	....
"	95.20	2.13	....	....	....	......	......	.....	.....	....	.50	S. & loss. 2.17
"	89.75	8.90	.62	tr.	....	......	......	.....	.....	....	.70	Mn. tr.
Cabin Creek......	91.87	6.60	tr.	....	....	.41	.05	Comb'nd .15	.....	tr.	.34	....
Cachiyuyal	93.92	4.93	.39	....	....	.08	......	.....	.20	....	....	Ca. Mg... .30

Loss.	Undet.	Total.	Sp. Gr.	Analyst.	Reference.
.....	.....	99.70	7.02-7.11	C. U. Shepard	1869, A. J. S. (2), XLVII, 233 234
.....	.....	100.58	7.746	J. Fahrenhorst	1900, Meteoreisen-Studien,XI, A.N.H. XV,371
.....	.....	99.96	....	J. Auerbach.........	1864, Sitz. Wien Akad., XLIX (2), 497
.....	.....	99.95	7.63	H. L. McIlwain.....	1888, A. J. S. (3), XXXVI, 277
.....	.....	99.54	....	O. W. Huntington...	1894, Proc. Am. Acad. Arts & Sci., XXIX, 253
.....	.....	100.38	7.74	W. Eberhard.......	1855, Ann. Chem. Pharm., XCVI, 286-289
.....	.....	100.00	....	W. P. Riddell.......	1860, Trans. St. Louis Acad. (1), 623
.....	.....	99.76	....	J. E. Whitfield......	1904, Proc. Rochester Acad. Sci., IV, 148
.....	.....	99.83	7.7	J. M. Davison.......	Same
.....	.....	100.38	....	L. Fletcher.........	1887, Min. Mag., VII, 125

the three alloys—kamacite, taenite, and plessite—are usually present. Among accessory constituents, troilite and schreibersite predominate. These are often in the form of nodules of appreciable size.

OCTAHEDRITES.

Loss.	Undet.	Total.	Sp. Gr.	Analyst.	Reference.
.....	.....	100.09	7.589	C. U. Shepard, Jr....	1876, A. J. S. (3), XII, 119
.....	.....	99.86	7.89	R. B. Riggs	1887, Bull. U. S. Geol. Sur. VIII, 94 97
.....	.....	100.26	7.75	A. A. Koch.........	1903, Bull. Geol. Soc. Amer. XIV, 104
.....	.....	100.00	....	M. A. Gobel	1874, Bull. St. Petersburg Akad. XIX, 544-54
.....	.....	100.45	....	F. F. Sharpless	1896, Amer. Geol. XVIII, 270
.....	.....	99.80	6.50-7.50	C. U. Shepard	1839, A. J. S. (1), XXXVI, 81 84
.....	.....	99.77	7.06	W. G. Owens.......	1892, A. J. S. (3), XLIII, 423-424
.....	.....	100.43	....	C. H. Rockwell	1844, A. J. S. (1), XLVI, 402
.....	.....	100.00	....	C. U. Shepard......	1847, A. J. S. (2), IV, 77-78
.....	.....	99.97	7.72	W. S. Clark	1852, Metallic Meteorites, 61-62
.....	.....	99.42	7.837	J. E. Whitfield......	1887, A. J. S. (3), XXXIII, 500
.....	.....	99.82	....	J. Domeyko	1875, Comptes Rendus, LXXXI, 597

Name.	Fe.	Ni.	Co.	Cu.	Cr.	P.	S.	C.	Si.	Cl.	In-sol.	Miscellaneous.
Caperr	89.87	9.33	.53	tr.	tr.	.24	......	.....	.....	....	....	
Cape York	90.14	8.18	.54	....	....	.18	.19	.15	.....	....		
" ....⅜	91.31	7.94	.53	.02	....	.19	.01	.04				
Carthage	89.47	7.72	.25	....	....	.09	.40	.....	.60	tr.	....	1.19
Casas Grandes....	95.13	4.38	.27	tr.	....	.24	......	tr.	.....			
"	92.66	7.26	.94	...	.03	.18	.02	.....			....	Chromite ..03
Chilkoot	92.56	7.11	.12	tr.	....	.12	.04	tr.				
Chulafinnee.......	91.61	7.37	.50	....	....	.17	......					
Cleveland.......⅜	89.59	8.79	.67	.12	....	.32	.006	.....				
Colfax.........²⁄₂	88.45	10.31	.57	.04	....	.19	.09	.....	.02			
"	88.05	10.37	.68	.04	....	.21	.08	.....	.02			
Coopertown.......	89.59	9.12	.35	tr.	....	.04	......					
Costilla..........	91.65	7.71	.44	....	....	.10	High. .26	.....				
Dalton	94.66	4.80	.34	....	tr.	tr.					....	Mn. tr.
Denton County....	94.02	5.43	tr.	....	.....	......	.....		.....	....	.33	....
"	92.10	7.53	tr.	....	....	.001	.....					
Descubridora	89.51	8.05	1.94	....	....	......	.45				....	P.Cr. and loss. .05
"	90.09	9.07		.24	....	......					.66	....
Elbogen	97.50	2.50	.....	....	....	......						
"	87.50	8.75	.....	....	tr.	......					....	Mn. tr.
"	88.23	8.52	1.85	....	....	......	tr.		.....		....	P. Fe. Ni. 2.21 Mg.28, Mn.tr.
"	89.90	8.43	.76	....	....	......					....	
"	94.69	2.47	.61	....	.12	......					....	Al. 19, Mn. .88
El Capitan.......	90.51	8.40	1.59	.05	....	.24	tr.		.....		....	
Fort Pierre	94.29	7.19	.60	....	.....	......	tr.		.....		....	Ca. 35, Mg. .65
"	90.76	7.61	.89	tr.	....	tr.	......				.05	
Franceville	91.10	8.06		....	....	......			tr.		....	*Pt. tr.
Frankfort	90.58	8.53	.36	tr.	....	.05	......				....	
Glorieta	88.76	9.86	.51	.03	....	.18	.01	.....	.04	......	....	Zn. .03, Mn.tr.

*Schreibersite, .84; Graphite, tr.; Silicate, tr.

Loss.	Undet.	Total.	Sp. Gr.	Analyst.	Reference.
.....	.....	99.97	7.86	L. Fletcher	1899, Min. Mag. XII, 167–170
.....	.....	99.38	.....	J. K. Phelps	1898, Northward Over the Great Ice, (2) 600
.....	.....	100.04	.....	J. E. Whitfield	" " " " 602
.....	.....	99.72	7.48–7.50	E. Boricky	1866, Neues Jahrb, 808–810
.....	.....	100.02	.....	W. Tassin	1902, Proc. U. S. Nat. Mus. XXV, 71
.....	.....	101.12	7.885	Cohen & Hildebrand	1903, Mitt. Nat. Ver. f. Neuvorp. u. Rügen, XXXV, 13
.05	.....	100.00	7.76	....	1905, Label, State Mining Bureau Collection, San Francisco, California
.....	.....	99.65	.....	J. B. Mackintosh	1880, A. J. S. (3), XX, 74
.....	.....	99.496	7.521	F. A. Genth	1886, Proc. Phila. Acad. Sci. 366–368
.....	.....	99.67	.....	S. W. Cramer	1890, Trans. N. Y. Acad. Sci. IX, 197–198
.....	.....	99.45	.....	L. G. Eakins	1890, A. J. S. (3), XXXIX, 395–396
.....	.....	99.10	7.85	J. L. Smith	1861, A. J. S. (2) XXXI, 266
.....	.....	100.16	.....	L. G. Eakins	1895, Proc. Colo. Sci. Soc.
.....	.....	99.80	7.986	C. U. Shepard, Jr.	1883, A. J. S. (3), XXVI, 338
.....	.....	99.78	7.67	W. P. Riddell	1860, Trans. St. Louis Acad. I, 623
.....	.....	99.63	7.42	A. Madelung	1863, Buchner, Meteoriten, 193
.....	.....	100.00	7.38	P. Murphy	1875, Neues Jahrb, 26
.....	.....	100.00	7.609	J. B. Mackintosh	1887, A. J. S. (3), XXXIII, 235
.....	.....	100.00	7.80–7.83	M. H. Klaproth	1815, Beit. Mineralkörper, VI, 306–308
.....	.....	98.10	7.76	J. F. John	1821, Jour. Chem. Phys. XXXII, 253–261
.....	.....	100.00	7.74–7.87	J. J. Berzelius	1834, Ann. Phys. Chem. XXXIII, 135–137
.06	.....	99.00	7.78	A. Wehrle	1863, Buchner, Meteoriten, 151–152
.....	.....	99.94	.....	P. A. v. Holger	" " " "
.....	.....	99.80	.....	H. N. Stokes	1895, A. J. S. (3), I, 252–254
.....	.....	102.48	7.73	H. A. Prout	1860, Trans. St. Louis Acad. I, 711–712
.....	.....	99.31	7.74	A. Madelung	1863, Buchner, Meteoriten, 197
.....	.....	100.00	7.87	J. M. Davison	1902, Proc. Roch. Aca. Sci. IV, 75–78
.....	.....	99.52	7.69	J. L. Smith	1870, A. J. S. (2), XLIX, 331
.....	.....	99.42	.....	L. G. Eakins	1885, Proc. Colo. Sci. Soc. II, 14

Name.	Fe.	Ni.	Co.	Cu.	Cr.	P.	S.	C.	Si.	Cl.	In-sol.	Miscellaneous.
Glorieta	88.81	7.28	.17	....	....	.12	......	....	....	....	....	....
"	87.93	11.15	33	....	....	.36	......	....	....	....	....	....
Guilford County	92.75	3.15	tr.	....	....	......	......	....	....	....	....	$Fe_2O_3 + FeO$ .75
Hopewell Mounds	95.20	4.64	.40	.04	....	.07	.13	....	....	....	....	Mn. tr., Sn. tr.
Hopper	90.54	7.70	.94	....	....	.13	......	....	.04	.35	....	....
Hraschina	96.50	3.50	.....								....	
"	83.29	11.84	1.26						.68	...		Mn. .64, Mg. .48 / K. .43, Al. 1.38
"	89.78	8.88	.67									
Ivanpah	94.98	4.52	.....			.07	......		.10		....	
Joel's Iron	90.45	8.80	.54	tr.	....	.26	......	tr.	....		....	
Joe Wright	91.22	8.62*		....	....	.16	......		....			
Juncal	92.03	7.00	.62	....	....	.21						
Kenton County	91.59	7.65	.84	tr.	....	tr.	tr.	.12				
Kokstad	91.21	8.01	.63	.02	....	.22	tr.	.03	......	.05	....	
La Caille ⅔	92.50	5.90	tr.	....	tr.	......	......		.90		....	
" ⅔	89.63	9.83	.....			......	......	.12	....	....		Insol. & loss. .42
Lenarto	85.04	8.12	3.59	....		......			.01			Ca. 1.63, Al. .77 / Mn. .61, Mg. .23.
"	90.90	8.50	.665	.002	....							
"	90.15	6.55	.50	.08	....	......		.48	.....	.....	1.23	Mn .15, Sn .08
"	90.88	8.45	.67	....								
"	91.50	8.58	.....	tr.		......					.30	....
Losttown	95.76	3.66	tr.	....	tr.	......					.58	Ca. tr.
Luis Lopez	91.31	8.17	.16	....	....	.33	.01	.01	tr.	.....	....	....
Marshall County	90.12	8.72	.32	tr.	....	.10						
Matatiela	92.20	7.30	.67	.03	....	.19	.03	.08	......	.03	....	
Mazapil	91.26	7.84	.65	....	....	.30	......					
Merceditas	92.38	7.33	.61	.02	....	.08	.07	.....	....	....	.02	....
Misteca	86.86	9.92	.74	....	....	.07	.55	.....	....		.97	....
Morito	95.01	4.22	.51	tr.	....	.08	.....	....	....	....	....	....

By diff.

Loss.	Undet.	Total.	Sp. Gr.	Analyst.	Reference.
.....	......	96.38	7.1	C. U. Shepard	1885, A. J. S. (3), XXIX, 469
.....	......	99.77	7.66	J. B. Mackintosh....	1885, A. J. S. (3), XXX, 238
.....	......	96.65	7.67	C. U. Shepard	1841, A. J. S. (1), XL, 369-370
.....	......	100.48	....	H. W. Nichols......	1902, Field Col. Mus. Pub. Geol. Ser. I, 308
.....	......	99.70	....	F. P. Venable	1890, A. J. S. (3), XL, 163
.....	......	100.00	7.73-7.80	M. H. Klaproth.....	1807, Beit. Mineralkörper, IV, 99-101
.....	......	100.00	7.82	P. A. v. Holger	1830, Beit. u. vor. Ett. Zeit. f. Phys. u. Math. VII, 2, 129-149
.....	......	99.33	7.785	A. Wehrle	1852, Clark, Metallic Meteorites, 42-44
.....	......	99.67	7.65	C. U. Shepard	1880, A. J. S. (3), XIX. 381-382
.....	......	100.05	7.863-7.958	L. Fletcher........	1889, Min. Mag. VIII, 264
.....	......	100.00	....	J. B. Mackintosh....	1886, A. J. S. (3), XXXI, 462
.....	......	99.86	....	A. A. Damour	1868, Comptes Rendus, LXVI, 569-571
.....	......	100.20	....	J. M. Davison	1892, A. J. S. (3), XLIV, 164
.....	......	100.17	7.7876	Fahrenhorst........	1900, Ann. S. Afr. Mus. II, 14
.....	......	99.30	7.43	L. E. Rivot	1854, Ann. Mines (5), VI, 554-555
.....	......	100.00	7.64	J. Boussingault	1872, Comptes Rendus, LXXIV, 1287-1289
.....	......	100.00	....	P. A. v. Holger	1830, Beit. u. Ett. Zeit. f. Phys. u. Math. VII, 2, 129-149
.....	......	100.067	7.79	A. Wehrle	1841, Rammelsberg, Handwörterbuch, 423
.....	......	99.22	7.73	W. S. Clark	1852, Metallic Meteorites, 40
.....	......	100.00	7.98	A. Wehrle	"　　"　　"　　"
.....	......	100.38	7.73	J. Boussingault	1872, Comptes Rendus, LXXIV, 1288-1289
.....	......	100.00	....	C. U. Shepard	1869, A. J. S. (2), XLVII, 234
.....	......	99.99	....	Mariner & Hoskins .	1900, A. J. S. (4), IX, 284
.....	......	99.26	....	J. L. Smith.........	1860, A. J. S. (2), XXX, 240
.....	......	100.53	7.8084	J. Fahrenhorst......	1900, Ann. So. Afr. Mus. II, 17
.....	......	100.05	....	J. B. Mackintosh....	1887, A. J. S. (3), XXXIII, 225.
.....	......	100.51	....	J. Fahrenhorst......	1900, Meteoreisen-Studien, XI, A. N. H. XV, 380
.....	......	99.11	7.58	C. Bergeman	1857, Pogg. Ann. C. 246
.....	......	99.82	7.84	J. L. Smith	1871, A. J. S. (3), II, 335-338

Name.	Fe.	Ni.	Co.	Cu.	Cr.	P.	S.	C.	Si.	Cl.	In-sol.	Miscellaneous.
Murfreesboro	96.00	2.40										
Nejed	91.04	7.40	.66	tr.		.10	tr.					
Orange River	90.48	8.94			tr.							Chladnite .56 Schreib. .02
Petropawlowsk	97.29	2.07										
" 2/3	93.57	6.98										
Plymouth	88.67	8.55	.66	.24		1.25	.07					Graph. .11
Puquios	88.67	9.83	.71	.04		.17	.09	.04	tr.			
Red River	90.02	9.67										
"	90.91	8.46								.50		
Rhine Valley	88.85	9.07	.34			.27	.75					
Roebourne	90.91	8.33	.06			.16	tr.	tr.		.01		Mn. tr.
Rowton	91.05	9.08		tr.								
"	91.25	8.58	.37	tr.								
Ruff's Mountain	96.00	3.12										
"	90.95	6.01	tr.		tr.						2.35	Schreib. .50
Sacramento Mts	91.39	7.86	.52									
San Angelo	91.96	7.86	tr.	.04		.10	.03	tr.	.01			Mn. tr.
Schwetz	93.18	5.77	1.05								.10	
Seneca Falls	92.40	7.60			tr.	tr.	fr.					Mg. tr. Mn. ? Sn. tr.
Staunton	91.44	7.56	.61	.02		.07	.02	.14	.11	tr.		Sn. tr.
"	90.29	8.85	.49	.02	tr.	.24	.01	.18	.69	tr.		Sn. .005 Mn. tr.
" No. 1	88.71	10.16	.40	.003		.34	.02	.17	.07	.003		Sn. .002 Mn. tr.
" No. 2	88.36	10.24	.43	.003		.36	.008	.18	.06	.002		Sn. .002
" No. 3	89.01	9.96	.39	.003		.37	.03	.12	.06	.004		Sn. .003 Mn. tr.
" No. 7	89.85	7.56	.60	.06		.16	.01	.05	.05			O. 1.56
Surprise Springs	91.01	7.65	.89	.07	.04	.22	.08	.02		.02		
Thunda	91.54	8.49	.56	.02	tr.	.17	.02				.01	
Toluca	91.38	8.62										
"	90.40	5.02	.04			.16						P. Fe. Ni. 2.99 Mn. tr.

Loss.	Undet.	Total.	Sp. Gr.	Analyst.	Reference.
.....	1.60	100.00	....	G. Troost	1848, A. J. S. (2), V, 351–352
.....	.59	99.79	7.89	L. Fletcher.........	1887, Min. Mag. VII, 179–182
.....	.....	100.00	7.3	C. U. Shepard	1856, A. J. S. (2), XXI, 213
.....	.....	99.36	7.76	Sokolowsky	1841, Arch. Kunde Russ. I, 317
.....	.....	100.55	....	Iwanow............	1841, Arch. Kunde Russ. I, 723–725
.....	.....	99.55	....	J. M. Davison.......	1895, A. J. S. (3), XLIX, 53–55
.....	.....	99.55	7.93	L. G. Eakins	1890, A. J. S. (3), XL, 226
.31	.....	100.00	7.54	C. U. Shepard	1829, A. J. S. (1), XVI, 217–219
.....	.....	99.87	7.40–7.82	B. Silliman, Jr. & T. S. Hunt.	1846, A. J. S. (2), II, 372–374
.....	.....	99.31	....	W. S. Chapman	1900, Ann. Rep. So. Aust. Sch. Mines, 227–228
.....	.....	100.00	....	Mariner & Hoskins .	1898, A. J. S. (4), V, 136
.....	.....	100.13	....	W. Flight	1882, Phil. Trans. 894–896
.....	.....	100.20	....	"	" " "
.....	.....	99.121	7.01–7.10	C. U. Shepard	1850, Proc. A. A. A. S. III, 152–154
.....	.....	99.81	....	Boecking	1856, Neues Jahrbuch, 51
.....	.....	100.00	7.7	Mariner & Hoskins .	1898, A. J. S. (4), V, 272
.....	.....	99.77	....	J. E. Whitfield......	1897, A. J. S. (4), III, 66
.....	.....	100.10	7.77	C. Rammelsberg....	1851, Ann. Phys. Chem. LXXXIV, 153–154
.....	.....	100.00	....	C. U. Shepard	1853, A. J. S. (2), XV, 366
.....	.....	99.97	7.69	J. P. Santos.........	1878, A. J. S. (3), XV, 337–338
.....	.....	100.175	....	J. W. Mallet........	1887, A. J. S. (3), XXXIII, 59
.....	.....	99.878	7.85	"	1871, A. J. S. (3), II, 13
.....	.....	99.345	7.86	"	" " "
.....	.....	99.95	7.84	"	" " "
.....	.....	99.90	....	J. E. Whitfield......	1903, A. J. S. (4), XV, 469–471
.....	.....	100.00	7.7308	E. Cohen..........	1900, Mitt. Nat. Ver. f. Neu Vorp. u. Rügen, 32
.....	.....	100.81	....	J. Fahrenhorst	1900, Meteoreisen-Studien, XI, A. N. H. XV, 382
.....	.....	100.00	7.72	Berthier...........	1853, A. J. S. (3), XV, 20
.....	.....	99.72	....	E. Uricoechea	1854, Jour. Prakt. Chem. LXIII, 317–318

Name.	Fe.	Ni.	Co.	C..	Cr.	P.	S.	C.	Si.	Cl.	In-sol.	Miscellaneous.
Toluca	90.37	7.79	.....	....	....	......	.....	....	.....	.....	1.91	....
"	90.72	8.49	.44	....	....	.18	......	.....	.25	....	....	....
"	87.88	8.86	.89	....	....	.86	......	.....			....	Graph.. 1.24
"	87.89	9.06	1.07	tr.	....	.62	......	.....			....	*C. Graph. .22
"	88.29	8.90	1.04	....	....	.78	......					....
"	90.08	7.10	.....	....	....	......	1.24	.....	....	....	....	....
"	90.43	7.62	.72	....	....	.15	.03	.....			....	†Cu. & Sn. .03
"	87.09	9.80	.77	.01	....	......	.79	.....	....	.....	.02	Schreib.... .73
"	89.07	7.29	.98	tr.	....	......	.85	.....	....	.....	.04	Mn. tr. Fe. S. tr.
"	90.13	7.24		....	....	.38	......				.22	....
"	91.89	6.32	1.58	....								Mn. tr.
(Los Reyes)	90.56	7.71	1.07	.14	....	.24	.03	.01	.01	.....	....	Mn. tr.
Tonganoxie	91.18	7.93	.39	tr.	....	.10	......					
Toubil	95.18	3.38	.14	....	....	.05	......	.12	.08	.04	....	Mn. .09 As. .02 Mg. .03 Ca. .21
Trenton	91.03	7.20	.53	tr.	....	.14	......	.....	.....	.....	.45	....
"	89.22	10.79	tr.	....	....	.69	......					
Victoria	91.33	8.83	.49	....								
Welland	91.17	8.54	.06	....	....	......	.07	.....				
Werchne Udinsk	91.02	7.31	.70	.13	....	.07	......	.03	....	.....	....	Mg. .03 Fe. Ni. P. .12
Wooster	93.61	6.01	.73	tr.	....	.13	......	....	.....	.....	....	Mn. tr.

*P. Fe. Ni., .34; Mn., .20; Sn., tr.
†Graph., etc., .34: P., Fe., Ni., .56.

Loss.	Undet.	Total.	Sp. Gr.	Analyst.	Reference.
.....	.....	100.07	....	W. J. Taylor........	1856, Proc. Phil. Aca. Sci. VIII, 3
...:	.....	100.46	....	"	1856, A. J. S. (2), XXII, 374–376
.....	.....	98.73	....	E. Pugh...........	1856, Ann. Chem. u. Pharm. XCVIII, 383-386
.....	.....	99.40	....	"	" " " "
....	.....	99.00	....	"	" " " "
.....	.....	98.42	....	"	" " " "
.....	.....	99.88	....	"	" " " "
.....	.....	99.21	....	Boecking...........	Neues Jahrbuch, 304
.....	.....	99.20	....	"	" "
2.03	.....	100.00	....	H. B. Nason........	1857, Jour. Prakt. Chem. LXXI, 123
....	.....	99.79	....	C. H. L. v. Babo	1863, Buchner, Meteoriten, 141
.....	.....	99.85	....	H. W. Nichols......	1902, Pub. Field Col. Mus., Geol. Ser, I 308
....	.....	99.60	7.45	E. H. S. Bailey	1891, A. J. S. (3), XLII, 386
.....	.....	99.34	....	J. Antipoff..........	1898, Bull. St. Petersburg Acad. Sci. V, 9, 91–103
.....	.....	99.35	7.82	J. L. Smith	1869, A. J. S. (2), XLVII, 271
.....	.....	100.70	7.33	G. Bode............	1869, Ann. Rep. Smith. Inst. 417–419
.....	.....	100.65	7.78	A. P. Coleman......	1887, Proc. and Trans. Roy. Soc. Can., IV, 97
.....	.....	99.84	7.87	J. M. Davison.......	1890, Proc. Roch. Acad. Sci. I, 87
.....	.....	99.41	....	H. Laspeyres.......	1895, Zeit. Kryst. XXIV, 494
.....	.....	100.48	7.90	J. L. Smith	1864, A. J. S. (2), XXXVIII, 385-386

4. FINE OCTAHEDRITES.

Width of lamellæ 0.4-0.2 mm. The nickel-cobalt content ranges between 8 and 10½ per cent. The fields are usually equal in amount to the lamellæ and contain minute shining flakes, probably of taenite.

a. PRAMBANAN

Name.	Fe.	Ni.	Co.	Cu.	Cr.	P.	S.	C.	Si.	Cl.	In-sol.	Miscellaneous.
Augustinowka	91.91	7.70	.25	....	....	......	......	....	....	....	....	....
Bella Roca........	91.48	7.92	.22	....	....	.21	.21	.06	....	....	....	....
"	89.68	9.78	.55	.02	tr.	.31	.05	....	....	....	....	....
Bethany (Mukerop) ..	91.07	8.18	.63	.03	.02	.06	.04	.01	....	....	....	....
" " .	92.29	7.77	.57	....	....	.06	......	....	....	....	....	Cu. C. Cr. Cl. S. .10
" " ..	90.96	8.19	.46	.04	....	.18	tr.	.02	....	.01	.01	....
" " ..	91.37	7.97	.50	.02	.04	.03	.02	.05	....	....	....	....
" (Lion River)..	93.30	6.70	....	....	....	tr.	tr.	....	....	....	....	K₂O tr. Sn. tr.
" " ..	92.06	7.79	.69	.03	.01	.05	.10	....	....	....	....	....
Boogaldi	91.13	8.05	.48	.28	....	......	......	....	....	....	.04	....
Bridgewater	88.90	9.94	.76	.35	....	......	......	....	.02	....	....	....
Bückeberg........	90.95	8.01		....	....	.64						
"	92.45	7.55	.83	.02	.01	.12	.01	.01	....	.02	....	....
Cambria ....½	94.88	5.69	....	....	....	......	......	....	....	....	....	....
"	92.58	5.71	....	tr.	....	......	......	....	....	....	1.40	As. tr.
"	89.06	10.65	.08	.04	....	......	.17	....	....	....	....	....
Charlotte	91.15	8.01	.72	.06	....	......	......	....	....	....	....	....
Chupaderos.......	90.23	8.76	1.21	....	....	tr.	......	....	....	....	....	....
"	88.78	9.80	.81	.02	tr.	.13	.13	....	....	....	....	....
Cuernavaca.......	88.98	10.30	....	....	....	......	......	....	....	....	....	....
"	89.70	8.76	1.19	.05	....	.33	.12	....	....	....	....	....
Grand Rapids.....	94.54	3.82	.40	....	....	......	......	....	....	....	.12	....
"	88.71	10.69	....	.07	....	.26	.03	.06	....	....	....	Mg. ...... .02 Graph... .07
Hassi Jekna.......	91.32	5.88	.81	tr.	....	......	tr.	....	....	....	1.04	....
Jamestown........	90.24	9.75	....	tr.	....	.05	......	....	....	....	....	

Cohen divides the fine octahedrites into two groups, the Prambanan group and the Rodeo group. The Prambanan group includes the greater number. They have a fairly uniform composition. Accessory constituents are usually present, but not in large quantity.

Group.

Loss.	Undet.	Total.	Sp. Gr.	Analyst.	References.
....	....	99.86	....	W. F. Alexejew	1893, Verh. Russ. Min. Ges. II, 30, 470
....	....	100.10	....	J. E. Whitfield......	1889, A. J. S. (3), XXXVII, 440
....	....	100.39	7.8244	Knauer	1905, Meteoritenkunde, III, 377
....	....	100.04	7.8408	J. Fahrenhorst......	1900, Ann. S. Afr. Mus. II, 28
....	....	100.79	7.8408	"	" " "
....	....	99.89	....	O. Hildebrand......	1902, Jb. d. Ver. f. Vaterl. Naturk. Württemberg, LVIII, 292–306
....	....	100.00	7.783	Krupp Lab.........	1902, Jb. d. Ver. f. Vaterl. Naturk. Württemberg, LVIII, 292–306
....	....	100.00	7.45	C. U. Shepard......	1853, A. J. S. (2), XV, 1–4
....	....	100.73	....	Sjöström & Fahrenhorst.	1897, Meteoreisen-Studien, V, A. N. H. XII, 43
....	....	99.98	7.85	A. Liversidge.......	1902, Jour. Roy. Soc. N. S. W. XXXVI, 341–359
....	....	99.97	6.617	F. P. Venable	1890, A. J. S. (3), XL, 312–313
....	....	99.60	7.12	Wohler & Wicke...	1863, Göttingen Nach. 364-367
....	....	101.01	....	J. Fahrenhorst......	1900, Meteoreisen-Studien, XI, A. N. H. XV, 367
....	....	100.57	7.52	D. Olmsted, Jr......	1845, A. J. S. (1), XLVIII, 388-392
....	....	99.69	....	B. Silliman, Jr., & T. S. Hunt.	1846, A. J. S. (2), II, 374–376
....	....	100.00	....	C. Rammelsberg ...	1870, Ber. Berl. Akad. 444
....	....	99.94	7.717	J. L. Smith	1875, A. J. S. (3), X, 351
....	....	100.20	....	Cohen & Weinschenk	1891, Meteoreisen-Studien, VI, A. N. H. VI, 147–148
....	tr.	99.67	....	O. Bürger..........	1905, Meteoritenkunde, III, 354
....	....	99.28	7.725	J. E. Whitfield	1902, Proc. Roch. Aca. Sci. IV, 79–88
....	....	100.15	7.748	O. Hildebrand......	1902, Mitt. d. Nat. Ver. f. Neu. Vorp. u. Rügen, XXXIV, 2
....	....	98.88	....	F. W. Taylor.......	1884, A. J. S. (3), XXVIII, 300
....	....	99.91	7.87	R. B. Riggs	1885, A. J. S. (3), XXX, 3·2
....	....	99.05	7.67	S. Meunier.........	1892, Comptes Rendus, CXV, 531–533
....	....	100.04	....	O. W. Huntington ..	1890, Proc. Am. Acad. (2), XVII, 229–232

Name.	Fe.	Ni.	Co.	Cu.	Cr.	P.	S.	C.	Si.	Cl.	In-sol.	Miscellaneous.
Jewel Hill	91.12	7.82	.43	tr.	....	.08						
Lagrange	91.21	7.81	.25	tr.	....	.05						
"	91.92	7.61	.62	.01	.02	.03	.02					
Mantos Blancos	90.77	8.83	.55	tr.	....	.10						
Moonbi	91.35	7.89	.56	tr.	tr.	.22		.07	.04			Sn. tr.
Prambanan	96.71	2.86										
"	94.36	5.37										
"	88.60	11.20				.20		.004	tr.			MgO. tr.
"	90.03	9.39	.97			.16						
"	94.38	4.70				.53						
Putnam County	89.52	8.82	tr.									Sn. P. S. Mg. Ca......1.66
"	90.28	7.89	.79	.07	.17	.11	.25					
Russel Gulch	90.61	7.84	.78	tr.	....	.02						
"	90.65	7.87	.01									Insol.Si,Sch.Cr. .95 Sn. .02
St. Genevieve.	91.58	7.98	.29			.20	tr.		.02			
Smith's Mountain	90.68	9.07		.11	....	.14						
"	90.88	8.02	.50	.03	....	.03						
Thurlow	89.17	9.92	1.04			.25	.05					Cu. & Cr., tr.
Varas	91.28	8.00	.44	tr.	....	.05						
Yanhuitlan	96.58	1.83						tr.	.01			CaO..... .61 Al₂O₃ .... .61
"	91.87	7.36	.65	.02	.01	.09	.02					

Loss.	Undet.	Total.	Sp. Gr.	Analyst.	Reference.
.....	.....	99.45	....	J. L. Smith.........	1860, A. J. S (2), XXX, 240
.....	.....	99.32	7.89	J. L. Smith.........	1861, A. J. S. (2), XXXI, 265–266
.....	.....	100.23	....	O. Bürger.........	1905, Meteoritenkunde, III, 358
.....	.....	100.25	7.904	L. Fletcher.........	1889, Min. Mag. VIII, 258
.....	.....	100.13	7.83	J. C. H. Mingaye ...	1893, Jour. Roy. Soc. N. S. W. XXVII, 82–83
.....	.....	99.57	7.48	M. Van der Boom Mesch ...	1866, Archives Neerl. I, 468
.....	.....	99.73	7.83	E. H. von Baumhauer	"　　"　　"　　"
.....	.....	100.004	....	Vlaanderen	1867, Nat. Tij. Ned. Ind., XXIX, 268–270
.....	.....	100.55	....	O. Sjöström	1897, Meteoreisen-Studien, A. N. H. XII, 42–62
.....	.....	99.61	....	De Jong	1904, Javabode, July 12, 5
.....	.....	100.00	7.69	C. U. Shepard	1854, A. J. S. (2), XVII, 331–332
.....	.....	99.56	....	Knauer & Bürger...	1905, Meteoritenkunde, III, 345
.....	.....	99.25	7.72	J. L. Smith.........	1866, A. J. S. (2), XLII, 218–219
.....	.....	99.50	7.692	C. T. Jackson	1867, A. J. S. (2), XLIII, 281
.....	.....	100.07	....	J. E. Whitfield	1901, Proc. Roch. Acad. Sci. IV, 65–66
.....	.....	100.00	....	F. A. Genth.........	1877, A. J. S. (3), XIII, 214
.....	.....	99.46	7.78	J. L. Smith.........	"　　"　　"
.....	.....	100.43	....	O. Bürger.........	1905, Meteoritenkunde, III, 379
.....	.....	99.77	7.863	L. Fletcher.........	1889, Min. Mag. VIII, 259
.36	.....	100.00	7.827	L. R. DeLoza.......	1876, Proc. Phila. Acad. Sci., 126
.....	.....	100.02	....	O. Bürger.........	1905, Meteoritenkunde, III, 320

b. RODEO GROUP

The nickel-cobalt content is somewhat higher than in the Pram-

RODEO

Name.	Fe.	Ni.	Co.	Cu.	Cr.	P.	S.	C.	Si.	Cl.	In-sol.	Miscellaneous.
Alt-Biela	85.34	12.89	.41	....	....	.39	.06	.02	....	....	.86	....
Bear Creek	83.89	14.06	.83	tr.	....	.21	......	....	....	....	....	....
Quesa	87.97	10.75	1.07	.04	....	.19	tr.	....	....	....	....	....
Rodeo	86.95	11.27	1.20	.01	.03	.25	.01	....	....	....	.07	....
"	89.84	8.79	.28	.07	....	.80	.02	.09	....	....	....	....

5. FINEST OCTAHEDRITES.

Width of lamellæ not exceeding 0.2 mm. The nickel-cobalt content lies, as a rule, between 10 and 15 per cent. Plessite strongly developed. Cohen divides the class into two groups, the Salt River group and the Tazewell group.

SALT RIVER

Name.	Fe.	Ni.	Co.	Cu.	Cr.	P.	S.	C.	Si.	Cl.	In-sol.	Miscellaneous.
Bacubirito	88.94	6.98	.21	....	....	.15	.005	....	tr.	....	....	....
"	89.54	9.40	.98	.02	.02	.12	.02	.01	....	.02	....	Chromite, .01
Ballinoo	89.34	9.87	.60	.06	....	.48	.03	.02	....	....	....	....
"	89.91	8.85	.74	tr.	....	.50	tr.	tr.	tr.	....	....	....
Butler ....½	89.12	10.02	.26	.01	....	.12	......	....	....	....	....	....
Salt River	90.74	9.36	....	....	....	......	tr.	....	....	....	.26	Mg. Na., tr.
" .....½	90.89	8.70	.85	.04	....	.34	tr.	.02	....	....	....	....

banan group, so that the group is a transition to the finest octahedrites.

Group.

Loss.	Undet.	Total.	Sp. Gr.	Analyst.	Reference.
....	.....	99.97	7.525	M. Neff & A. Stocky	1899, Prog.d.Böhm Gym.in Mahr-Ostrow
....	.....	98.99	....	J. L. Smith	1867, A. J. S. (2), XLIII, 280
....	.....	100.02	....	J. Fahrenhorst......	1900, Meteoreisen-Studien, XI, A. N. H. XV, 379
....	.....	99.79	....	O. Bürger..........	1905, Meteoritenkunde, III, 299
....	.....	99.89	....	H. W. Nichols......	1905, Field Col. Mus. Geol. Ser. III, 4

a.   Salt River Group.

The content of nickel-cobalt is lower than in the Tazewell group, not exceeding 10½ per cent. Plessite predominates as compared with the Tazewell group. Schreibersite is common in numerous small, elongated individuals.

Group.

Loss.	Undet.	Total.	Sp. Gr.	Analyst.	Reference.
....	.....	96.285	7.69	J. E. Whitfield	1902, Proc. Roch. Acad. Sci. IV, 74
....	.....	100.14	7.59	Cohen & Hildebrand	1903, Mitt. N. Ver. f. Neu Vorp. u. Rügen, XXXV, 13
....	.....	100.40	7.8432	O. Sjöström........	1898, Ber. Berlin Akad., 19–22
....	.....	100.00	7.8	Mariner & Hoskins.	1898, A. J. S. (4), V, 137
....	.....	99.53	7.72	J. L. Smith........	1877, A. J. S. (3), XIII, 213
....	.....	100.36	....	W. H. Brewer......	1851, Proc. A. A. A. Sci. IV, 36-38
....	.....	100.84	7.6648	J. Fahrenhorst	1900, Meteoreisen-Studien, X, A. N. H. XV, 76

b. TAZEWELL GROUP.

This group includes the octahedrites having the highest percent-

TAZEWELL

Name.	Fe.	Ni.	Co.	Cu.	Cr.	P.	S.	C.	Si.	Cl.	In-sol.	Miscellaneous.
Carlton..........	86.54	12.77	.63	.02	....	.16	.03	.11	....	....	....	....
Laurens County...	85.33	13.34	.87	....	....	.16	tr.	tr.	....	....	....	....
Mart............	89.68	9.20	.33	.04	tr.	.16	.02	....	....	....	....	....
Mungindi.........	90.31	8.23	1.36	....	....	.09	tr.	.01	tr.	....	....	....
"	87.96	10.99	.88	.08	.03	.17	.21	....	....	....	....	....
Narraburra Creek.....¼	88.60	9.74	.47	.01	....	.43	tr.	....	....	....	.72	Resinous matter .01
Tazewell........⅔	82.70	14.82	.46	.07	....	.18	.08	....	....	.02	....	Mg.0 .24, Si0₂ .65

c. COWRA AND VICTORIA WEST.

Cowra and Victoria West resemble the Salt River group in their predominance of plessite.   Their place among the octahedrites is not

COWRA AND

Name.	Fe.	Ni.	Co.	Cu.	Cr.	P.	S.	C.	Si.	Cl.	In-sol.	Miscellaneous.
Cowra ....⅔	85.26	13.23	1.02	.02	....	.22	.01	.03	.01	....	....	Sn.and Mn.,tr.
Victoria West.....	88.83	10.14	0.53	tr.	....	.28	......	....	....	....	....	....

B. HAMMOND OCTAHEDRITES.

These meteorites appear in section to be granular aggregates in which black particles and taenite-like lamellæ extend in directions parallel to octahedral planes.   They thus have resemblances to the octahedrites and form a transition to the ataxites.   The structure by which they are characterized may be either original or secondary.   If original, the structure has been produced by the separation of the nickel-rich alloy and black particles to form a web, the lines of which

HAMMOND

Name.	Fe.	Ni.	Co.	Cu.	Cr.	P.	S.	C.	Si.	Cl.	In-sol.	Miscellaneous.
Cacaria..........	87.38	12.06	.65	.02	.01	.22	.05	....	....	....	.09	SiO₂ ... .16
"	92.00	7.70	.54	.03	.01	.24	.06	....	....	....	....	....
Hammond........	89.78	7.65	1.32	tr.	....	.51	......	tr.	....	....	....	Sn. tr. SiO₂ .56
"	91.62	7.34	1.01	.04	.01	.52	.01	.06	....	.01	....	....
Reed City	89.39	8.18	....	....	....	......	....	....	....	....	....	....

age of nickel-cobalt.   It reaches 15 per cent. and more.   Taenite is strongly developed.

GROUP.

Loss.	Undet.	Total.	Sp. Gr.	Analyst.	Reference.
.....	......	100.26	7.95	L. G. Eakins.......	1890, A. J. S. (3), XL, 223–224
.....	......	99.70	....	J. B. Mackintosh....	1886, A. J. S. (3), XXXI, 463–465
.....	......	99.43	....	H. N. Stokes	1900, Proc. Wash. Acad. Sci. II, 53
.....	......	100.00	7.4	Mariner & Hoskins.	1898, A. J. S. (4), V, 139
.....	......	100.32	....	R. Knauer	1905, Meteoritenkunde, III, 269
.....	......	99.98	7.57	A. Liversidge.......	1903, Proc. Roy. Soc. N. S. W. XXXVII, 240
.....	......	99.22	7.89	J. L. Smith.........	1855, A. J. S. (2), XIX, 153

certain, however, and it seems desirable therefore to group them separately.   Their percentage of nickel-cobalt resembles that of the finest octahedrites, 11 to 15 per cent.

VICTORIA WEST.

Loss.	Undet.	Total.	Sp. Gr,	Analyst.	Reference.
.....	......	99.80	7.805	J. C. H. Mingaye ...	1904, Rec. Geol. Sur. N. S. W., VII, 31
.....	......	99.78	7.692	J. L. Smith.........	1873, A. J. S. (3), V. 108

accord with octahedral planes.   In the meshes of this web the nickel-poor remainder is deposited as a homogeneous, granular aggregate.   If the structure is secondary, it may be explained by supposing that a normal octahedrite was somewhat softened by heat, so as to destroy the lamellar structure in part, after which solidification took place.   If this latter be the correct explanation, the softening was carried farther in Hammond than in Cacaria and Reed City.

OCTAHEDRITES.

Loss.	Undet.	Total.	Sp. Gr.	Analyst.	Reference.
.....	......	100.64	7.7070	J. Fahrenhorst......	1900, Meteoreisen-Studien, XI, A. N. H., XV, 362–363
.....	......	100.58	....	"	Same
.....	......	99.82	7.601–7.703	Fisher & Allmendinger	1887, A. J. S. (3), XXXIV, 383
.....	......	100.62	7.288–7.506	J. Fahrenhorst......	1900, Meteoreisen-Studien, XI, A. N. H., XV, 356
.....	......	97.57	7.6	J. E. Whitfield	1903, Jour. Geol., XI, 233

C. BRECCIATED OCTAHEDRITES.

In these, as in the brecciated hexahedrites, the mass appears to be

BRECCIATED

Name.	Fe.	Ni.	Co.	Cu.	Cr.	P.	S.	C.	Si.	Cl.	In-sol.	Miscellaneous.
Barranca Blanca..	91.50	8.01	.65	tr.	....	.15	.13	....	....	....	.03	....
N'Goureyma......	89.28	9.26	.60	.04	.11	.05	.77	.04	....	....	.24	Ce...... .01
"	91.99	7.15	tr.	....	....	......	......	....	....	....	....	Chromite .09 Fe.S.... .05 Graphite, etc. ..17
Santa Rosa .....⅔	91.46	7.72	....	....	....	......	......	....	....	....	.28	....
"	92.30	6.52	.78	.02	tr.	.36	.04	.18	....	....	....	....
Tula	93.50	2.50	....	....	....	......	......	....	....	....	....	Sn. tr. Schreibersite.. .90
"	96.40	2.63	....	....	....	......	......	....	....	....	....	Sn. .07 Sch. .90
Zacatecas........	89.84	5.96	.62	tr.	....	......	.13	....	....	....	3.08	Mg. tr.
"	90.91	5.65	.42	....	....	.23	.07	......	.50	....	2.17	....
"	91.30	5.82	.41	tr.	....	.25	......	....	....	....	2.19	Mg., tr.
"	92.09	5.98	.91	....	.74	1.02	......	....	....	....	.04	....

made up of numerous individuals, the direction of whose lamellæ differs in the individual grains.

Octahedrites.

Loss.	Undet.	Total.	Sp. Gr.	Analyst.	Reference.
.....	......	100.47	7.823	L. Fletcher	1889, Min. Mag., VIII, 263
.....	......	100.49	7.6722	E. Cohen	1901, Mitt. Nat. Ver. f. Neu. Vorp. u. Rügen, XXXIII, 14
.....	......	99.36	7.31	S. Meunier........	1901, Compt. Rendus, CXXXII, 444
.....	......	99.46	7.30-7.60	Rivero and Boussingault ...	1824, Ann. Phys. Chem., XXV, 438–443
.....	......	100.20	7.6896	O. Sjöström.......	1899, Meteoreisen-Studien, VIII, A. N. H., XIV, 138
.....	......	96.90	7.332	W. Haidinger	1861, A. J. S. (2), XXXII, 144
.....	......	100.00	....	J. Auerbach	1863, Neues Jahrb., 362
.....	......	99.63	7.20	H. Müller..........	1860, Jour. f. prakt. Chemie, LXXIX, 25
.....	......	99.95	7.625	"	" " " "
.....	......	99.97	7.50	"	" " " "
.....	......	100.78	....	E. Cohen	1897, Meteoreisen-Studien, V, A. N. H., XII, 51

III.   ATAXITES.

These iron meteorites are characterized by a fine granular to compact structure throughout.   They show no evidence of the cubic cleavage and Neumann lines which characterize the hexahedrites, nor of the lamellar structure, octahedrally arranged, of the octahedrites. The individual grains are in some cases visible to the naked eye, but for the most part are of microscopic or sub-microscopic dimensions. In some occur peculiar streaks which seem to have crystallographic arrangement, but their exact relations have not been determined. These form a special group, which, while not ataxites in the strictest sense of the term, may be included among them for present purposes.   The ataxites show the greatest variation among all iron meteorites in their nickel-cobalt content.   This varies from 6 to 16 per cent, and in the doubtful Oktibbeha to 63 per cent.   Two general

SIRATIK

Name.	Fe.	Ni.	Co.	Cu.	Cr.	P.	S.	C.	Si.	Cl.	In-sol.	Miscellaneous.
Campo del Cielo .. (Wöhler's Iron.)	92.33	7.38	....	....	.....	......	....	....	.....	....	P. Fe. Ni. .42 Sn .... .03
....	89 22	9.51	....	....	.20	......	tr.	....	.....	....	Schreibersite. .06 C. etc.... .24
....	94.25	5.11	.57	.03	.03	.18	.05	.....	.....	tr.	....	....
Cincinnati	94.47	5.43	.68	.01	....	.05	.05	.....	....	....	....	....
Locust Grove	94.30	5.57	.64	tr.	....	.18	.05	.02	.....	.01	....	....
San Francisco del Mezquital	93.38	5.89	.39	....	....	.23	......	....	....	....	....	....
" "	93.36	5.46	.87	.03	....	.16	.15	.....	....	....	....	....
Siratik (Senegal)..	94.07	5.21	.77	.01	....	.26	.04	.01	.....	....	....	....

subdivisions may be made of the ataxites, according as they are nickel-poor or nickel-rich.   Transitions occur between these, but a general grouping is practicable.   Accessory constituents are not usually abundant in the ataxites, and when occurring are of small dimensions as a rule.

A. NICKEL-POOR ATAXITES.

The nickel-cobalt content lies between 6 and 7 per cent, the composition thus corresponding to that of kamacite.   The structure is, as a rule, plainly granular, seldom compact, the size of the grains reaching 0.75 mm.

1. SIRATIK GROUP.

An etched surface appears rough through the presence of irregularly arranged depressions, due perhaps to the solution of some accessory constituent, such as troilite or schreibersite.   The smaller the depressions the more plainly the boundaries of the grains appear. The latter range from 0.33 to 0.75 mm. in dimension.

Group

Loss.	Unde .	Total.	Sp. Gr.	Analyst.	Reference.
. . . . .	. . . . . .	100.16	7.547	N. S. Manross.	1853, A. J. S. (2), XV, 22
. . . . .	. . . . . .	99.23	7.85	C. Martius	Ann. Chem. u. Pharm., CXV, 92
. . . . .	. . . . . .	100.28	7.7679	O. Sjöström	1898, Meteoreisen-Studien, VIII, A. N. H., XIII, 124
. . . . .	. . . . . .	100.69	7.6895	"	1898, Ber. Berlin Akad., 428–430
. . . . .	. . . . . .	100.77	7.7083	"	1897, Ber. Berlin Akad., 76–81
. . . . .	. . . . . .	99.89	7.83	A. A. Damour.	1868, Comptes Rendus, LXVI, 573–574
. . . . .	. . . . . .	100.03	7.7687	J. Fahrenhorst	1900, Meteoreisen-Studien, XI, A. N. H., XV, 365
. . . . .	. . . . . .	100.37	7.7752	O. Sjöström	1898, Meteoreisen-Studien, VIII, A. N. H., XIII. 131

2. NEDAGOLLA GROUP.

Both granular and compact irons occur in this group.   They lack the rough appearance of the Siratik group on etched surfaces.   The

NEDAGOLLA

Name.	Fe.	Ni.	Co.	Cu.	Cr.	P.	S.	C.	Si.	Cl.	In-sol.	Miscellaneous.
Chesterville.......	95.00	5.00	....	....	tr.	tr.	......	....	....	....	....	....
"	93.15	5.82	.73	....	....	.34	......	....	....	....	....	....
"	93.80	5.50	.75	.02	tr.	.34	.03	.02	....	....	....	....
Forsyth County...	94.90	4.18	.33	....	....	tr.	.22	....	....	....	....	....
(Compact portion.).	94.03	5.55	.53	.02	....	.23	.03	.02	....	tr.	....	....
(Granular portion).	94.18	5.56	.60	.02	....	.19	.05	.04	....	.17	....	....
Nedagolla	92.61	6.20	.49	tr.	....	.02	.05	....	.25	....	....	....
Nenntmansdorf ...	94.50	5.31	....	....	....	......	......	....	....	....	....	....
" ...	93.04	6.16	....	....	....	.22	......	....	....	....	....	....
" ...	94.33	5.48	.71	....	....	.29	......	....	....	....	....	....
Primitiva	94.72	4.72	.71	tr.	....	.18	.02	.03	....	....	....	....
Rasgata	90.76	7.87	....	....	....	......	......	....	....	....	....	P. Fe. Ni. .37
"	92.35	6.71	.25	tr.	....	.35	tr.	....	....	....	....	Silicates . .08
"	92.81	6.70	.64	.01	tr.	.28	.08	.19	....	....	....	Sn....... tr.

3. RAFRUTI GROUP.

The members of this group resemble the granular members of the

RAFRUTI

Name.	Fe.	Ni.	Co.	Cu.	Cr	P.	S.	C.	Si.	Cl.	In-sol.	Miscellaneous.
Illinois Gulch......	92.51	6.70	.16	....	....	.62	......	.01	tr.	....	....	....
" ....⅔	86.77	12.67	.81	.02	.01	.08	tr.	....	....	....	....	....
Rafrüti	89.87	9.54	.61	.03	.01	.06	.11	.18	....	....	....	....

size of the grains in the granular members is generally less than 0.5 mm., rarely 0.75 mm.   No granular structure is visible, even on strong magnification, in the compact members.   Chesterville and Rasgata are rich in rhabdite.

GROUP.

Loss.	Undet.	Total.	Sp. Gr.	Analyst.	Reference
.....	......	100.00	7.82	C. U. Shepard	1849, A. J. S. (2), VII, 449
.....	......	100.04	....	O. Sjöström	1897, Meteoreisen-Studien, V, A. N. H., XII, 47
.....	......	100.46	7.8209	"	1898, Meteoreisen-Studien, VIII, A. N. H., XIV, 150
.....	......	99.63	....	E. A. de Schweinitz	1896, A. J. S. (4), I, 208–209
.....	......	100.41	7.4954	O. Sjöström........	1897, Ber. Berlin Akad., 386–396
.....	......	100.81	7.3357	"	" " "
.....	......	99.62	7.8613	"	1897. Meteoreisen-Studien, VI, A. N. H., XII, 121
.....	......	99.81	....	G. E. Lichtenberger	1873, Sitz. Isis. p. 4, Dresden
.....	......	99.42	6.21	E. Geinitz	1876, Neues Jahrb., 609
.....	......	100.81	7.8241	E. Cohen...........	1897, Meteoreisen-Studien, V, A. N. H., XII, 42
.....	......	100.38	....	O. Sjöström........	1897, Meteoreisen-Studien, VI, A. N. H., XII, 123
.....	......	98.63	7.6	Rivero and Boussingault....	1824, Ann. Chem. Phys., XXV, 442–443
.....	......	100.11	7.33–7.77	F. Wöhler..........	1852, Ann. Chem. Pharm., LXXXII, 243–248
.....	......	100.71	7.654	O. Sjöström	1898, Meteoreisen-Studien, VIII, A. N. H., XIII, 143

Nedagolla group, but have an essentially higher nickel-cobalt content, and thus form a transition to the nickel-rich ataxites.

GROUP.

Loss.	Undet.	Total.	Sp. Gr.	Analyst.	Reference.
.....	......	100.00	7.7	Mariner and Hoskins	1900, A. J. S. (4), IX, 201–202
.....	......	100.36	7.8329	J. Fahrenhorst......	1900, Meteoreisen-Studien, XI, A. N. H., XV, 353
.....	......	100.41	7.596	Cohen and Hildebrand.....	1902, Mitt. Nat. Ver. f. Neu. Vorp. u. Rügen, XXXIV, 87

B. NICKEL-RICH ATAXITES.

These ataxites are fine-grained to compact, and acquire, as a rule, on weak etching, a characteristic varnish-like luster. Stronger etching produces a dull surface, having a peculiar velvety sheen. The

SMITHLAND

Name.	Fe.	Ni.	Co.	Cu.	Cr.	P.	S.	C.	Si.	Cl.	In-sol.	Miscellaneous.
Babb's Mill .... (Troost Iron)	85.30	14.70	.....	....	....	......	......	....	....	.....	....	Al. Mg. Ca., tr.
"	87.16	9.76	.....	....	....	......	......	....	....	.....	....	....
"	80.59	17.10	2.04	....	....	......	......	....	tr.	.....	.12	Mn. ...... .tr. P. Fe. Ni. .12
"	81.54	17.74	1.26	....	....	.11	......	....	....	.....	....	P. Fe. Ni. .05
"	81.45	17.30	1.67	.03	.03	.12	.01	.07	.....	.....	....	....
(Blake Iron)	91.42	7.95	.....	....	....	......	......	....	....	.....	....	....
"	86.30	12.58	1.66	....	....	......	......	....	....	.....	....	....
"	88.23	11.01	.72	....	.02	tr.	tr.	.03	.....	.01	....	....
"	88.41	11.09	.66	....	.02	tr.	tr.	.03	.....	.02	....	....
Botetourt	85.88	18.23	.....	....	....	......	......	....	....	.....	....	............,
Deep Springs.....	87.01	11.69	.79	....	....	.04	......	....	.53	.39	....	....
"	85.99	13.44	.70	.03	.03	.06	......	.02	.....	.02	....	....
Dehesa...........	86.20	14.20	.....	....	....	......	......	....	....	.....	....	....
Linville	84.56	14.95	.33	....	....	tr.	.12	tr.	.....	.....	....	....
"	83.13	16.32	.76	.02	....	.23	.02	.11	.....	.....	....	....
Morradal	79.67	18.77	1.18	.06	.06	.18	.27	....	.....	.....	....	....
Smithland	82.83	16.42	.94	....	.06	.09	.17	....	.....	.....	....	....
Weaver	80.78	17.92	.84	....	....	.12	.15	....	.....	.....	.15	....

nickel-cobalt content lies, for the most part, between 14 and 20 per cent, though it drops to 12 and rises to 26½ per cent.

1. SMITHLAND GROUP.

The nickel-cobalt content does not exceed 20 per cent.

GROUP.

Loss.	Undet.	Total.	Sp. Gr.	Analyst.	Reference.
.....	......	100.00	7.548	C. U. Shepard	1847, A. J. S. (2) IV, 76–77
.....	......	96.92	....	G. Troost	1845, A. J. S. (1), XLIX, 342–344*
.....	......	99.85	7.839	W. S. Clark........	1852, Metallic Meteorites, 65–66
.....	......	100.70	7.7948	E. Cohen	1892, Meteoreisen-Studien, II, A. N. H., VII, 147, 148
.....	......	100.68	....	J. Fahrenhorst	1900, Meteoreisen-Studien, X, A. N. H., XV, 93
.....	......	99.37	7.858	W. P. Blake........	1886, A. J. S. (3), XXXI, 44
.....	......	100.54	....	Cohen and Weinschenk.....	1891, Meteoreisen-Studien, I, A. N. H., VI. 142–143
.....	......	100.02	....	J. Fahrenhorst	1900, Meteoreison-Studien, X, A. N. H., XV, 93
.....	......	100.23	....	"	Same
.....	......	104.11	8.186	O. Sjöström........	1898, Meteoreisen-Studien, VII, A. N. H., XIII, 49
.....	......	100.45	....	F. P. Venable......	1890, A. J. S. (3), XL, 162
.....	......	100.29	7.4538	J. Fahrenhorst	1900, Meteoreisen-Studien, XI, A. N. H., XV, 355
.....	......	100.40	7.8892	J. Domeyko........	1879, Mineralojia, Santiago
.....	......	99.96	....	J. E. Whitfield	1888, A. J. S. (3), XXXVI, 276
.....	......	100.59	7.4727	O. Sjöström........	1898, Meteoreisen-Studien, VIII, A. N. H., XIII, 147
.....	......	100.19	7.8543	"	1898, Videnskabsselskabets Skrifter (1), VII, 11
.....	......	100.51	7.7115	"	1898, Meteoreisen-Studien, VII, A. N. H., XIII, 47
.....	......	99.96	7.12	Lindner............	1904, Sitzb. K. Preus. Akad. der Wiss. XXXII

*As recalculated by Cohen, Meteoritenkunde, Heft III, p. 104.

2. CRISTOBAL GROUP.

The nickel-cobalt content exceeds 20 per cent.

CRISTOBAL

Name.	Fe.	Ni.	Co.	Cu.	Cr.	P.	S.	C.	Si.	Cl.	In-sol.	Miscellaneous.
Limestone Creek..	65.18	27.71	.....	.....	.....	.....	.....	.....	.....	.....	....	....
" " ..	66.56	24.71	.....	.....	.....	.....	4.0	.....	.....	1.48	....	Cr.&Mn. 3.24
" " ..	83.57	12.67	.....	.....	.....	.....	.....	.....	.....	.91	....	FeS$_2$ 2.39
" " ..	65.03	29.99	1.48	....	.....	.19	.....	.....	.....	.....	....	....
San Cristobal	73.72	25.60	1.0	....	.....	.18	.....	.....	.....	.....	....	....

3. OKTIBBEHA.

The meteoric origin of Oktibbeha is doubtful, on account of its

OKTIBBEHA

Name.	Fe.	Ni.	Co.	Cu.	Cr.	P.	S.	C.	Si.	Cl.	In-sol.	Miscellaneous.
Oktibbeha	37.69	59.69	.40	.90	....	.10	.....	.....	.12	.....	....	Al...... .20 Ca...... .09
"	37.24	62.01	.72	.28	....	.15	.....	.....	.....	.....	....	....

C.   ATAXITES WITH ACCESSORY FORSTERITE.

The accessory occurrence of forsterite is characteristic. It forms about five per cent of the mass, occurring in small spheroidal grains or elongated aggregates of grains, and is accompanied by some plagioclase.   In nickel-cobalt content the metallic portion of the meteorite

ATAXITES WITH

Name.	Fe.	Ni.	Co.	Cu.	Cr.	P.	S.	C.	Si.O$_2$	Cl.	In-sol.	Miscellaneous.
Tucson	85.54	8.55	.61	.03	....	.12	.....	.....	3.02	.....	....	MgO 2.04 Cr$_2$O$_3$.21 Al$_2$O$_3$, tr.
"	83.55	9.20	.39	.01	.17	.13	.....	.....	3.01	.....	....	Labradorite. 1.05 CaO .51 MgO2.26*
(Carleton Iron)......	81.56	9.17	.44	.08	....	.49	.....	.....	3.63	.....	....	FeO .12 CaO 1.16 MgO 2.43 0 .61
"	84.56	8.89	1.36	.03	.02	.16	tr.	.04	1.72	.04	....	MgO .. .59 Chrys. res. 3.68
(Ainsa Iron)	84.60	9.24	.95	.02	.02	.17	.01	.04	1.76	.04	....	MgO .. .51 Chrys.res.. 3.39

* K$_2$O, .10; Na$_2$O, .17.

GROUP.

Loss.	Undet.	Total.	Sp. Gr.	Analyst.	Reference.
.....	......	92.89	5.75	C. T. Jackson.......	1838, A. J. S. (1), XXXIV, 335
.....	......	99.99	5.75–6.40–6.50	"	" " · " "
.46	......	100.00	6.82	A. A. Hayes........	1845, A. J. S. (1), XLVIII, 153
.....	......	96.89	....	R. Knauer	1905, Meteoritenkunde, III, 131
.....	......	100.50	7.8593	"	1899, Ber. Berlin Akad., 607–608

anomalous composition. It may however for the present be included among meteorites.

GROUP.

Loss.	Undet.	Total.	Sp. Gr.	Analyst.	Reference.
.....	......	99.19	6.854	W. J. Taylor	1857, A. J. S. (2), XXIV, 294
.....	......	100.40	....	E. Cohen	1892, Meteoreisen-Studien, II, A. N. H., VII, 146

lies between the nickel-rich and nickel-poor ataxites. On etching, irregularly shaped areas appear, 0.2–2 cm. in area, which under the microscope have a spotted look and are generally bordered, as are most of the silicate grains, by narrow, zigzag bands the nature of which cannot be further determined.

ACCESSORY FORSTERITE.

Loss.	Undet.	Total.	Sp. Gr.	Analyst.	Reference.
.....	......	100.12	6.52–7.13	J. L. Smith	1855, A. J. S. (2), XIX, 161–162
.....	......	100.55	....	F. A. Genth........	1855, A. J. S. (2), XX, 119–120
.....	......	99.69	7.29	G. J. Brush........	1863, A. J. S. (2), XXXVI, 153
.....	......	101.09	7.2248	J. Fahrenhorst......	1900, Cohen-Festschrift, Greifswald, 39
.....	......	100.75	....	"	" " "

D. ATAXITES WITH CUBIC STREAKS.

Upon etching appear bands or spots which seem to be oriented according to cubic faces, and which according to the position of the plates toward impinging light appear brighter or darker than the principal mass of the nickel-iron without a structural distinction being discernible.   In one position the reflection of the whole face is plainly uniform.   On weak etching appears, as a rule, a characteristic luster.

ATAXITES WITH

Name.	Fe.	Ni.	Co.	Cu.	Cr.	P.	S.	C.	Si.	Cl.	In-sol.	Miscellaneous.
Cape of Good Hope	78.90	15.28	1.0									Ca. 1.41, Al. .16 Mg. .15, Mn. 1.76 Graphite....1.34
"	85.61	12.27	.89									
"	81.20	15.03	2.56	tr.		.09	tr.				.95	Sn....... tr.
"	81.30	15.23	2.01	tr.		.08	tr.					P. Fe. Ni. .88 Sn....... tr.
"	82.77	14.32	2.52	tr.		.26						
"	82.87	15.67	.95	.03	.04	.09		.03		.01		
Iquique	83.83	15.86	.19			.05					.07	
"	83.49	15.41	.94	.02	tr.	.07	.02	.03				
Kokomo	87.02	12.29	.65	tr.		.02						
"	83.24	15.76	1.07	.01		.08	tr.					
Shingle Springs	88.02	8.88									3.5	
"	80.74	15.73										Sn....... .01 P. etc....3.52
"	81.48	17.17	.60		.02	.31	.01	.07	.03			Ca. .16, Al. .09 Mg. .01, K. .03
"	82.21	16.69	.65	.02	.02	.34	.05	.03				
Ternera	83.02	16.22	1.63	tr.								
"	82.17	16.22	1.42			.11	.13					

On strong etching the surface becomes dull with a peculiar velvet sheen.
No cleavage has been observed.   On the other hand, a certain orientation of similarly situated particles is indicated by the appearance in
reflected light.   The structure of the nickel-iron is compact; the content in nickel plus cobalt 16–17 per cent.   Except for the etch bands,
the members of this group are similar in chemical composition and
luster to the etched faces of members of the Morradal group.

Cubic Streaks.

Loss.	Undet.	Total.	Sp. Gr.	Analyst.	Reference.
.....	......	100.00	7.544	V. Holger..........	1830, Zeit f. Phys. u. Math., VIII, 279–284
.....	......	98.77	7.66	A. Wehrle.........	1835, Zeit. f. Phys. u. Math. (2), III, 222–229
.....	......	99.89	6.63–7.94	E. Uricoechea......	1854, Ann. Chem. u. Pharm., XCI, 252
.....	......	99.50	7.60	M. Böcking	1855, Ann. Chem. u. Pharm., XCVI, 243–246
.....	......	99.87	7.71	Baumhauer and Seelheim ...	1867, Arch. Neerland, II, 376–384
.....	......	99.69	7.8543	J. Fahrenhorst	1900, Meteoreisen-Studien, X, A. N. H., XV, 87
.....	......	100.00	7.925	C. Rammelsberg....	1873, Fest. Ges. Natur. Freunde, Berlin, 37
.....	......	99.98	7.8334	O. Sjöström........	1898, Meteoreisen-Studien, VIII, A. N. H., XIII, 153
.....	......	99.98	7.821	J. L. Smith.........	1874, A. J. S. (3), VII, 392
.....	......	100.16	7.8606	O. Sjöström........	1898, Meteoreisen-Studien, VIII, A. N. H., XIII, 118–158
.....	....	100.00	7.80	C. U. Shepard......	1872, A. J. S. (3), III, 438
.....	......	100.00	7.9053	C. T. Jackson.......	1872, A. J. S. (3), IV, 495
.....	......	99.98	7.875–8.024	F. A. Cairns........	1873, A. J. S. (3), V, 21
.....	......	100.01	7.8943	O. Sjöström........	1898, Meteoreisen-Studien, IX, A. N. H., XIII, 479–480
.....	......	100.87	7.694	E. Weinschenk.....	1892, A. J. S. (3), XLIII, 425
.....	......	100.05	....	Lindner...........	1904, Ber. Berlin Akad., 151

ANALYSES OF OCCLUDED GASES.

The occluded gases of nine iron meteorites have been determined. These are here shown.

Name.	Vols.	H.	CO₂	CO.	N.	CH₄	Analyst.	Reference.
Charlotte	2.20	71.40	13.30	15.30	.....	.....	A. W. Wright..	1876, A. J. S. (3), XI, 257
Cranbourne	3.59	45.79	0.12	31.88	17.66	4.55	W. Flight	1882, Ph. Tr. Roy. Soc. London, 893 896
Lenarto ..:....	2.85	85.68	.....	4.46	9.86	.....	Th. Graham ...	1866, Proc. Roy. Soc., London, XV, 502–503
Magura	47.13	18.19	12.56	67.71	1.54	.....	A. W. Wright..	1876, op. cit.
Red River.....	1.29	76.79	8.59	14.62	.....	....	A. W. Wright..	1876, op. cit.
Rowton	6.38	77.78	5.15	7.34	9.72	.....	W. Flight	1882, op. cit.
Shingle Springs	0.97	68.81	13.64	12.47	5.08	.....	A. W. Wright..	1876, op. cit.
Staunton	3.17	35.83	9.75	38.33	16.09	.....	J. W. Mallet ...	1871, Proc. Roy. Soc. London, XX, 365–370
Tazewell	3.17	42.66	14.40	41.23	1.71	.....	A. W. Wright..	1876, op. cit.

DISCUSSION OF ANALYSES.

The most striking feature brought out by the analyses is the relation shown between chemical composition and structure. This seems to be definite and general. All the meteorites of a hexahedral structure have a nearly uniform composition, while among the octahedral meteorites, fineness of structure increases with increase of nickel. This conclusion can best be shown by obtaining the averages from the analyses of the different groups, omitting all obviously faulty analyses. The results thus obtained are as follows:

Class.	No. of Analyses.	Width of Lamellæ in Millimeters.	Per Cent Fe.
Hexahedrites..............	29	------	94.12
Coarsest Octahedrites....	12	+ 2.5	93.18
Coarse "	22	2.0–1.5	92.28
Medium "	88	1.0–0.5	90.64
Fine "	41	0.4–0.2	90.18
Finest "	13	0.2– —	88.51

It is worthy of note that these averages are not means between wide limits, but are derived from nearly uniform values. Practically all of the members of the classes conform in composition to the average. Were all the groups equally well known, it is probable, too, that the gradation of percentage of Fe would be even more uniform than here shown. The medium octahedrites, for example, while numerous, have been as a whole imperfectly analyzed. Moreover, some of the meteorites classed as medium octahedrites, which are characterized by low percentage of iron, such as Algoma and Glorieta Mountain, have width of lamellæ such as to place them near if not in the fine octahedrites.

The apparent conclusion from the above results is, that the content of nickel influences the structure. It may also account for the change from a hexahedral to an octahedral structure, since the irons with a hexahedral structure have the lowest per cent of nickel. So constant and definite does this relation hold, that given a certain structure the per cent of nickel can probably be stated more accurately by this principle than it has been determined in some analyses. The per cent of nickel in iron meteorites as a whole, as shown by the reliable analyses, lies between five and twenty-six per cent. An exception to the latter figure may be found in the quoted analyses of Limestone Creek, but of this unfortunately no complete analysis exists. The somewhat doubtful Oktibbeha is also an exception, its percentage of nickel reaching sixty per cent. Cobalt in the iron meteorites rarely exceeds one per cent. No constant relation in amount appears to exist between it and nickel, although perhaps as a rule it is higher with higher nickel. Copper is doubtless, as claimed by Smith, a constant ingredient of iron meteorites. It is usually only a few hundredths of one per cent in amount, but may reach a few tenths. Chromium is shown by the analyses to be a frequent though not constant ingredient in minute quantities. In many cases it is probably present as daubreelite, but also, as suggested by Cohen, it may occur as an element alloyed with nickel-iron. Reports of the presence of manganese and tin are so frequent as to leave little doubt that they occur in many iron meteorites, perhaps alloyed as metals. The presence of platinum and iridium has been proved by Davison in Coahuila and Franceville, and doubtless could be found to exist in more meteorites if proper search were made. Gold was reported in Boogaldi by Liversidge, but in so small a quantity as to make its determination as yet not quite positive. The presence of occluded gases has been determined in but few cases. The constant presence of phosphorus in iron meteorites is a feature shown by the analyses. Apparently no iron meteorite is lacking in this element altogether, and in amount and constancy it con-

siderably exceeds sulphur.   It probably occurs combined with nickel-
iron as phosphide.   Sulphur, though evident by its presence in many
meteorites as troilite, does not appear in large amounts in the analyses,
and does not seem to be so important or constant an ingredient as
phosphorus.   Carbon is probably more frequent in occurrence than
analyses usually show, since of twenty-eight iron meteorites investigated by Cohen for carbon all but one showed appreciable percentages,
ranging from .19 per cent to .012 per cent.*   The silicon reported in
the analyses is doubtless in some cases to be referred to silicate grains,
but in other cases may be free or combined with the iron as a silicide.
The analyses make plain the incompleteness of much of the work
which has been done hitherto.   There can be little doubt that complete analyses of iron meteorites should always show iron, nickel,
cobalt, copper, and phosphorus, and in most cases sulphur, carbon,
and silicon.   When considerable differences occur in the analyses of
the same meteorite, as, for instance, 2 per cent of nickel reported in
Burlington by Rockwell and nearly 9 per cent by Shepard, the difference
is probably not to be regarded as due to the meteorite, but to the
analyses.   In a substance made up of different alloys and accessory
minerals as are the iron meteorites, especially the octahedrites, there
can be no question that portions from different parts of the meteorite
would of necessity show unlike composition.   How wide these varia-
tions might legitimately be it is difficult to say, but some causes of
error may be suggested.   One of these is imperfect sampling.   The
proper method to secure material for mass analyses of an iron meteorite, especially if of octahedral structure, is to use dust obtained by
boring.   A mixture of the constituents of the meteorite is thus obtained
which insures a better representation of its composition than is possible
when only a fragment broken from some part of the surface is used.
Such a fragment may contain an excess of taenite, or be largely composed of some accessory mineral so as to be far from representing the
true constitution of the meteorite.   Yet the larger number of analyses
of iron meteorites have probably been made with fragments of this character, and the wonder is, not that they show so much variation, but that
they do not show more.   Meteorites also doubtless vary in their homo-
geneity, as shown especially by Canyon Diablo, in one portion of which
Moissan found 2.89 per cent of nickel, and in another, only one centimeter distant, 5.06 per cent.   In another piece of Canyon Diablo
two analyses made by the same analyst of material obtained at dis-
tances of one centimeter showed 1.17 per cent and 7.11 per cent of

* Meteoritenkunde, Heft II., p. 243.

nickel.* While few meteorites probably vary to this extent, such determinations show the need of as thorough sampling as possible if a mass analysis is to be made. Occasionally a marked variation in the analyses of a meteorite seems explicable only on the assumption that the material analyzed did not belong to that meteorite. Such, for instance, seems the most reasonable explanation for the percentage of nickel, 12.67 per cent, reported by Hayes for Limestone Creek, as compared with the percentages, 25–30 per cent, obtained by other analysts. Errors of this sort are obviously difficult to detect, and can only be surmised in extreme cases. Another and more serious cause of discrepancies in analyses is the imperfect separation by the analyst of nickel and cobalt from the iron. The methods for this separation are not altogether satisfactory, even at the present day, and in earlier years they were much less so. Consequently the results of the earlier analysts were for the most part too low in these ingredients. The determinations of specific gravity shown in the tables appear in some cases to have been equally open to sources of error with the analyses. It can easily be calculated that the specific gravity of an iron meteorite is likely to be between 7.6 and 7.9, since the specific gravity of pure iron, 7.85, will be increased by that of nickel, 8.8, according to the proportion of the latter. It will be decreased by accessory minerals, such as troilite, which has a specific gravity of 4.7, schreibersite, 6.5, graphite, 2.2, and oxidized ingredients. Any porosity of the meteorite will also lessen its specific gravity. It is obvious, therefore, that determinations of specific gravity made on small fragments can hardly represent that of the mass as a whole, since they may contain a disproportionate quantity of accessory ingredients or may be more oxidized than the main mass. It is hardly credible that porosity or accessory ingredients of a meteorite would in any case reduce its specific gravity below 7. Determinations below this figure, therefore, probably indicate that oxidized material was used. From the showing in the tables that large numbers of meteorites have practically similar composition, it is evident that similarity of composition cannot be used, as has often been done hitherto, to prove identity of origin of meteorites found at different places. This method at one time obtained considerable vogue. Dissimilarity of composition, on the other hand, as a rule indicates separate falls. The only marked exception to this rule seems to be furnished by the two masses of Babb's Mill, one of which shows about 11 per cent, the other about 17 per cent, of nickel. The only alternative supposition possible here

* C. R., 1893, cxvi., 290.

is that two ataxites fell at different times at one locality. In view of the small number of ataxites known, this seems less likely than to suppose that two masses of the same fall differed in composition. No other case of such marked difference is known. Differences of structure seem as a rule to be a better criterion for distinguishing meteorites than differences of composition. On the other hand, similarity of structure and composition together do not positively identify meteorites found at different places as belonging to one fall, since such similarities occur in meteorites seen to fall at widely different times and places. Of the nine iron meteorites seen to fall, four are medium octahedrites and have practically similar compositions. In correlating individual meteorites, therefore, all possible characters must be taken into consideration, including the circumstances of their find, the appearance of their exterior, the probable time elapsed since their fall, etc.

No attempt has been made by the writer at summation of the analyses here given, in order to determine the average composition of iron meteorites. Such a summation, if worthy of being performed at all, will be deferred until analyses of the iron-stone and stone meteorites are also at hand for comparison. This work the writer hopes to accomplish in the near future. It is obvious, however, from an inspection of the tables that the average percentage of iron in iron meteorites as a whole is not far from 91 per cent, while that of nickel closely approximates 7.50 per cent. It is doubtful if the average percentage of the remaining minor constituents can be learned by summation of existing analyses. Not only have these constituents in many cases not been determined, but also any slight error in analyses or sampling would double or multiple their percentage. A percentage of .4 of cobalt, for instance, as compared with .2, is within the limits of error of many analyses, yet one percentage is double that of the other. The same is true in much greater degree of determinations of the amount of copper and other constituents. Until a larger number of complete and accurate determinations are at hand, therefore, summations of these constituents seem to have little value. One point in the composition of iron meteorites which may or may not be of significance may be noted. Of the four constant metallic constituents, the most abundant, iron, has the lowest atomic weight, the next in quantity, nickel, is next higher, and so on for cobalt and copper. This gradation, using percentages common in iron meteorites, appears as follows:

	Iron.	Nickel.	Cobalt.	Copper.
Per cent in iron meteorites	90	9	0.9	0.02
Atomic weight	55.5	58.3	58.6	63.1

Field Museum of Natural History.

Publication 151.

Geological Series.   Vol. III, No. 9.

ANALYSES OF STONE METEORITES

COMPILED AND CLASSIFIED
BY

Oliver Cummings Farrington

Curator, Department of Geology

Chicago, U. S. A.

June 1, 1911.

ANALYSES OF STONE METEORITES COMPILED AND CLASSIFIED.

BY OLIVER CUMMINGS FARRINGTON.

The object of this publication is twofold: (1) To give a compilation of analyses of stone meteorites of the same nature as that already made by the author for iron meteorites.* (2) To use these analyses as a basis for the establishment of a quantitative classification. The plan on which the analyses have been collected for the first purpose has already been described in the introduction to the paper on Analyses of Iron Meteorites. The need of such a collection is due to the fact that as with the iron meteorites, the last extensive compilation of analyses of stone meteorites which was published was that of Wadsworth in 1884.† Since Wadsworth's compilation a number of excellent analyses have been made both of meteorites which have fallen since that time and of earlier ones, and the convenience of having these analyses grouped together for purposes of reference is obvious. The chief difference between the collection by the present writer of the analyses of the stone meteorites and that of the iron meteorites is that a more rigid selection of the analyses of the stone meteorites has been made. Only those analyses which gave satisfactory evidence of being thorough and complete have been admitted to the list. On the other hand tolerance has been exercised in the admission of analyses which might on the whole be complete although obviously containing minor errors. The greatest difficulty which has been encountered in including analyses in the collection has been that of obtaining mass analyses. It has been a common tendency of analysts of stone meteorites to give only analyses of separate portions. In order to combine the analyses of the separate portions into a mass analysis a reduction of all results to 100 is, of course, necessary. The results thus obtained probably often fail to accurately represent all the constituents of the meteorite, but on the

---

* Analyses of Iron Meteorites Compiled and Classified, Field Col. Mus. Pub. 1907, Geol. Ser., Vol. 3, pp. 59–110.

† Rocks of the Cordilleras; Mem. Mus. Comp. Zool. Cambridge, Mass., 1884, Vol. II, pt. 1, pp. XVI–XXXIII.

whole no serious error need be involved.   To confine reported analyses to those which were only stated in the mass form would reduce the number materially and fail to represent our true knowledge of the chemical composition of meteorites.

The second purpose for which the grouping of the analyses has been made was, as has been stated, to propose a quantitative classification. The principles of this classification are the same as those for terrestrial rocks proposed by Cross, Iddings, Pirsson, and Washington.*   It was suggested by Washington in his publication on the Chemical Analyses of Igneous Rocks and their Classification† that such a classification of meteorites be made, and the writer held a brief conference with Dr. Washington on the subject.   The need of such a classification of meteorites is, perhaps, even more acute than was the case with terrestrial rocks.   Of the various classifications of meteorites which have been proposed none can be considered quantitative.   The classification chiefly used for stone meteorites at the present time is that which has been gradually evolved through the labors of Rose, Tschermak, Cohen, and Brezina.   It is presented in its most complete form by Brezina in the Catalogue of the Ward-Coonley Collection of Meteorites.‡   As is well known, the groups of this classification are based primarily upon structure but also upon mineralogical characters.   The stones are first subdivided into achondrites, chondrites, and siderolites.   The achondrites are divided into a number of groups distinguished by mineralogical composition.   These include the eukrites, chladnites, howardites, etc.   Among the chondrites the subdivisions are based chiefly on color, the groups being designated as white, gray, black, intermediate, carbonaceous, etc., with additional divisions according to structure giving spherulitic and crystalline.   Other subdivisions are based upon the presence or absence of veins and breccia-like structure. Of these divisions, that according to color cannot be regarded as resting upon any important or fundamental character, although it finds some slight justification in the fact that the lighter-colored meteorites are likely to contain more enstatite than the darker ones.   Another weak feature of the classification in the view of the present writer is its failure to take account, in any definite way, of the metallic content of meteorites.   The metal of meteorites is an important feature which should serve as a distinguishing mark.

So far as the iron meteorites are concerned the present system of

* Quantitative Classification of Igneous Rocks, Chicago, 1903.
† U. S. Geological Survey, 1903, Prof. Pap. No. 14, pp. 9 and 61.
‡ Henry A. Ward, Chicago, 1904, pp. 97–101.

Brezina is quantitative, as the present writer has shown.* The metallic content of the stone meteorites, however, finds little recognition in the Brezina system.

It will be obvious that some modification of the Quantitative Classification of terrestrial rocks is necessary in order to fit it for use with meteorites. Among these one is due to the impossibility of using regional names for the nomenclature of orders, sections, etc., of meteorites. For this reason in designation of the subdivisions the writer has used only descriptive adjectives. A group name is given only to the last group, the subrang. This name is that of a meteorite as nearly representative in composition as possible, preference being given, where there is a choice of names, to the better known meteorites. Another modification of classification necessary has been on account of the abundance of metal in meteorites. This required the formation of several subclasses in the classes in which among terrestrial rocks but a single subclass exists. Two subclasses are thus required in Class IV and four in Class V. As no nomenclature was proposed by the authors of the Quantitative Classification which would be applicable to more than one subclass, it has been necessary for the writer to provide names for the additional subclasses. This has been done by coining adjective terms indicating the relative quantities of silicates and metal. The adjectives for the five subdivisions are: persilicic, dosilicic, silico-metallic, dometallic, and permetallic. As will be noted by consulting the tables, most meteorites fall outside of the groups of terrestrial rocks. The following groups are similar in meteorites and terrestrial rocks: Kedabdekase of terrestrial rocks corresponds to Juvinose of meteorites; Wehrlose to Udenose; Argeinose to Stawropolose; Maricose to Bishopvillose; and Websterose to Bustose. Some minerals not found in terrestrial rocks occur in meteorites. To these the writer has given the following abbreviations: troilite, *tr;* oldhamite, *oh;* nickel-iron, *nf.* As it is occasionally necessary to assume the presence of the molecule (Mg, Fe)O in meteorites, the name *femite* and abbreviation *mo* are proposed for it. The standard minerals assumed to be present in meteorites and their abbreviations are then as follows:

GROUP I: SALIC MINERALS

Quartz, SiO_2		Q
Zircon, $ZrO_2 . SiO_2$		Z
Orthoclase, $K_2O . Al_2O_3 . 6 SiO_2$	or	
Albite, $Na_2O . Al_2O_3 . 6 SiO_2$	ab	F
Anorthite, $CaO . Al_2O_3 . 2 SiO_2$	an	

* Field Col. Mus. Pub. 1907, Geol. Ser., Vol. 3, p. 108.

Leucite, $K_2 O . Al_2 O_3 . 4 Si O_2$.................... lc
Nephelite, $Na_2 O . Al_2 O_3 . 2 Si O_2$............... ne $\Big\}$ L
Kaliophilite, $K_2 O . Al_3 O_2 2 Si O_2$................. kp

GROUP II: FEMIC MINERALS

Acmite, $Na_2 O . Fe_2 O_3 . 4 Si O_2$.................. ac
Sodium metasilicate, $Na_2 O . Si O_2$................ ns
Potassium metasilicate, $K_2 O . Si O_2$.............. ks
Diopside, $Ca O . (Mg, Fe) O . 2 Si O_2$.............. di $\Big\}$ P
Wollastonite, $Ca O . Si O_2$...................... wo
Hypersthene, $(Mg, Fe) O . Si O_2$................. hy"
Olivine, $2 (Mg, Fe) O . Si O_2$..................... ol $\Big\}$ O
Akermanite, $4 Ca O . 3 Si O_2$..................... am
Magnetite, $Fe O . Fe_2 O_3$....................... m_t
Femite, $Mg, Fe O$............................. mo $\Big\}$ H
Chromite, $Fe O . Cr_2 O_3$........................ om
Hematite, $Fe_2 O_3$............................... hm
Ilmenite, $Fe O . Ti O_2$........................... il   T $\Big\}$ M
Apatite, $3 (3 Ca O . P_2 O_5) . Ca F_2$............... ap
Troilite, $Fe S$.............................. tr
Oldhamite, $Ca S$............................... oh $\Big\}$ A
Schreibersite, $(Fe, Ni)_3 P$....................... sc
Nickel-iron, Fe_n, Ni_m........................... nf

The methods of calculating the analyses of meteorites in order to determine their place in this classification are the same as those adopted for terrestrial rocks by the authors of the Quantitative Classification. These are given in detail in their publication. As it may be convenient, however, to have the quantitative classification of meteorites so far as possible complete in itself, so much of the methods of calculation as is deemed necessary is here repeated from the work of the authors of the Quantitative Classification.*

1. Determine the molecular proportions of the chemical components of a rock as expressed by the complete analysis, by dividing the percentage weights of each component by its molecular weight.

2. Before undertaking the distribution of the chemical components as mineral molecules, small amounts of $Mn O$ and $Ni O$ are to be united with $Fe O$, and of $Ba O$ and $Sr O$ with $Ca O$; of $Cr_2 O_3$ with $Fe_2 O_3$, unless these unusual components occur in sufficient amounts to make their calculation as special mineral molecules desirable.

3. Establish the fixed molecules by allotting:

a) to $Cr_2 O_3$, if present in notable amount, $Fe O$ to satisfy the ratio $Cr_2 O_3 : Fe O :: 1 : 1$ for chromite:

b) to $Ti O_2$ enough $Fe O$ to satisfy the ratio $Ti O_2 : Fe O :: 1 : 1$ for ilmenite. If there is excess of $Ti O_2$, allot to it equal $Ca O$ for titanite or perofskite according to available silica, to be determined later. If there is an excess of $Ti O_2$ it is to be calculated as rutile.

* *Loc. cit.* pp. 188–195.

c) to P_2O_5 allot enough Ca O to satisfy the ratio P_2O_5 : Ca O :: 1 : 3.33 for apatite. Allot F or Cl to satisfy Ca O = 0.33 P_2O_5;

d) to F not used in apatite allot Ca O to form fluorite, Ca O : F :: 1 : 2;

e) to Cl allot Na_2 O in the ratio Cl_2 : Na_2 (O) :: 1 : 1 for sodalite;

f) to SO_3 allot Na_2 O in proportion SO_3 : Na_2 O :: 1 : 1 for noselite;

g) to S allot Fe O in proportion S : Fe (O) :: 2 : 1 for pyrite;

h) to C O_2 in undecomposed rocks allot Ca O in the proportion 1 : 1 for calcite. CO_2 may occur in primary calcite and cancrinite. If these minerals are secondary, the CO_2 is to be neglected, since it is understood that analyses of decomposed rocks are not available for purposes of classification.

Having adjusted the minor, inflexible, molecules, there remain the more important but variable silicate molecules, which form the great part of the mineral composition, or *norm*, of most rocks.

4. To Al_2 O_3 are allotted all the K_2 O and Na_2 O not already disposed of, in the proportion of Al_2 O_3 : K_2 O + Na_2 O :: 1 : 1 for alkali feldspathic and feldspathoid (lenad) molecules.

5. With excess of Al_2 O_3, $(Al_2 O_3 > K_2 O + Na_2 O)$;

a) to extra Al_2 O_3 allot Ca O in proportion of Al_2 O_3 : Ca O :: 1 : 1 for anorthite molecules.

b) If there is further excess of Al_2 O_3 it is to be considered as corundum, Al_2 O_3.

6. With insufficient Al_2 O_3, $(Al_2 O_3 < K_2 O + Na_2 O)$;

a) Extra Na_2 O is alloted to Fe_2 O_3 in proportion Fe_2 O_3 : Na_2 O :: 1 : 1 for acmite molecules.

b) If there is still extra Na_2 O it is set aside for a metasilicate molecule (Na_2 Si O_3).

c) When there is an excess of K_2 O over Al_2 O_3 it is treated in the same manner. It is an extremely rare occurrence.

7. In working with reliable analyses in which Fe_2 O_3 and Fe O have been correctly determined:

a) To Fe_2 O_3 is allotted excess of Na_2 O under conditions 6, *a*).

b) To remaining Fe_2 O_3 is allotted available Fe O in equal proportions for magnetite.

c) If there is any excess of Fe_2 O_3 it is calculated as hematite.

Analyses in which all the iron has been determined in one form of oxidation, when it occurs in two, are of little value when considerable iron is present. When the amount of iron is very small the analyses may still be used as a means of classifying the rock. For this purpose all the iron, if given as ferric oxide, is to be calculated as Fe O, except that necessary to be allotted to Na_2 O for acmite, and then used as below.

8. *a*) Extra Ca O after the foregoing assignments is allotted to (Mg, Fe) O in proportion Ca O : (Mg, Fe) O :: 1 : 1 for diopside molecules.

In all molecules where (Mg, Fe) O is present, Mg O and Fe O are to be used in the same proportions in which they are found after Fe O has been allotted to the molecules previously mentioned. That is, they are to be introduced into diopside, hypersthene, and olivine with the same ratio between them.

b) If there is still an excess of Ca O it is to be set aside for calcium metasilicate (Ca Si O_3) or subsilicate (4 Ca O . 3 Si O_2), equivalent to wollastonite or akermanite. Such extra Ca O will in most cases actually enter garnet, an alferric mineral.

9. With insufficient Ca O, (Ca O < (Mg, Fe) O);

a) Extra (Mg, Fe) O is to be set aside for metasilicate or orthosilicate, hypersthene or olivine, according to the amount of Si O_2 present.

The allotment of Si O_2 to form silicates begins with the bases which occur with silica in but one proportion, and is carried on as follows:

10. To Zr O_2 allot Si O_2 in proportion of 1 : 1 for zircon.

11. To Ca O and $Al_2 O_3$ in anorthite is allotted equal Si O_2 to form Ca O.$Al_2 O_3$.2 Si O_2.

12. To Ca O and (Mg, Fe) O in diopside is allotted equal Si O_2 to form Ca O.(Mg, Fe) O.2 Si O_2.

13. To Na_2 O and $Fe_2 O_3$ in acmite is allotted Si O_2 to form Na_2 O.$Fe_2 O_3$.4 Si O_2.

14. To Na_2 O (or K_2 O) set aside for metasilicate molecules allot Si O_2 to form Na_2 O.Si O_2 or K_2 O.Si O_2.

15. To Na_2 O and $Al_2 O_3$ in sufficient amount to form with Na Cl sodalite, or with $Na_2 SO_4$ noselite, is allotted Si O_2 to satisfy the formulas : 3 (Na_2 O.$Al_2 O_3$ 2Si O_2).2 Na Cl, sodalite, 2 (Na_2 O.$Al_2 O_3$.2 Si O_2).$Na_2 SO_4$ noselite.

16. To Ca O set aside for wollastonite or akermanite is allotted tentatively Si O_2 to form wollastonite (Ca O.Si O_2).

17. To extra (Mg, Fe) O is allotted Si O_2 to form orthosilicate, olivine (2 (Mg, Fe) O.Si O_2).

18. To $Al_2 O_3$ and K_2 O + Na_2 O is allotted Si O_2 to make the polysilicates, orthoclase and albite, K_2 O.$Al_2 O_3$.6 Si O_2 and Na_2 O. $Al_2 O_3$.6 Si O_2.

a) If there is an excess of Si O_2 it is added to the orthosilicate of (Mg, Fe) O to raise it to the metasilicate (Mg, Fe) O.Si O_2. If Si O_2 is insufficient to convert all the olivine into hypersthene it is distributed according to the following equations:

$$x + y = \text{molecules of (Mg, Fe) O.}$$

$$x + \frac{y}{2} = \text{available Si } O_2.$$

where x = hypersthene, $\dfrac{y}{2}$ = olivine molecules.

b) Further excess of Si O_2 is to be allotted to Ti O_2 and Ca O to form titanite. These constituents remain as perofskite when there is no excess of Si O_2.

c) Further excess of Si O_2 is reckoned as quartz.

19. If there is insufficient Si O_2 to form polysilicate feldspar out of all the K_2 O and Na_2 O with $Al_2 O_3$:

a) To K_2 O.$Al_2 O_3$ is allotted tentatively enough Si O_2 to form polysilicate, orthoclase (K_2 O.$Al_2 O_3$.6 Si O_2) and the remaining Si O_2 is distributed between albite and nephelite molecules by means of the equations:

$$x + y = \text{molecules of } Na_2 \text{ O.}$$
$$6x + 2y = \text{available Si } O_2.$$

where x = albite, and y = nephelite molecules.

b) If the available Si O_2 in case 15, *a)* is insufficient to form nephelite with the Na_2 O, then enough Si O_2 is first allotted to the Na_2 O to form nephelite and the remaining Si O_2 is distributed between orthoclase and leucite molecules by means of the equations:

$$x + y = \text{molecules of } K_2 \text{ O.}$$
$$6x + 4y = \text{available Si } O_2.$$

where x = orthoclase, and y = leucite molecules.

20. If there is insufficient Si O_2 to form leucite and nephelite with olivine it is necessary to reduce a sufficient number of molecules to form the subsilicate akermanite, 4Ca O.3 Si O_2.

a) In case there is no wollastonite this is done after distributing SiO_2 tentatively to form leucite, nephelite, and olivine, and noting the deficit of SiO_2 by means of the equation:

$$y = \frac{1}{3} \text{ of the deficit of } SiO_2.$$

$$y = \text{molecules of akermanite. } (4\,CaO.3\,SiO_2).$$

CaO is to be taken from diopside, and the MgO and FeO so liberated are to be calculated as olivine.

b) In case an excess of CaO has been set aside for wollastonite this is first converted to akermanite by means of the equations:

$$y = \text{the deficit of } SiO_2.$$

$$y = \text{molecules of akermanite } (4\,CaO.3\,SiO_2).$$

c) If there are not sufficient molecules of wollastonite to satisfy the deficit of silica, recalculate the molecules of diopside and wollastonite so as to make olivine, diopside, and akermanite by means of the formulæ:

$$2x + 3y + \frac{z}{2} = \text{available } SiO_2.$$

$$x + 4y \quad = \text{molecules of } CaO.$$

$$x + z \quad\quad = \text{molecules of } MgO + FeO.$$

where x = molecules of new diopside, y = molecules of akermanite $(4\,CaO.3\,SiO_2)$, and z = molecules of olivine.

21. If there is still not enough SiO_2, all the CaO of the diopside and wollastonite must be calculated as akermanite, the $(Mg, Fe)O$ being reckoned as olivine and the K_2O distributed between leucite and kaliophilite by the equations:

$$x + y \quad = \text{molecules of } K_2O.$$

$$4x + 2y = \text{available } SiO_2.$$

where x is K_2O in leucite and y = K_2O in kaliophilite.

22. In case there is insufficient SiO_2 and an excess of Al_2O_3 and $(Mg, Fe)O$, which might form aluminum spinel, an alferric mineral, the excess of Al_2O_3 is to be calculated as corundum, and the uncombined $(Mg, Fe)O$ is to be estimated as femic minerals, being placed with the nonsilicate, mitic group, magnetite, ilmenite, etc.

GLOSSARY

A

Alkalicalcic. Having salic alkalies and salic lime present in equal or nearly equal amounts. $\dfrac{K_2 O' + Na_2 O'}{Ca O'} < \dfrac{5}{3} > \dfrac{3}{5}$.

C

Calcimiric. Equally calcic and miric, or nearly so. $\dfrac{Mg O + Fe O}{Ca O} < \dfrac{5}{3} > \dfrac{3}{5}$.

Class. Division of igneous rocks based on the relative proportions of salic and femic standard minerals.

D

Do- (or dom) Prefix indicating that one factor dominates over another within the ratios $\dfrac{7}{1}$ and $\dfrac{5}{3}$.

Docalcic. Dominantly calcic. Of salic minerals when Ca O' dominates over $K_2 O' + Na_2 O'$. $\dfrac{K_2 O' + Na_2 O'}{Ca O'} < \dfrac{3}{5} > \dfrac{1}{7}$. Of femic minerals when Ca O'' dominates over Mg O + Fe O. $\dfrac{Mg O + Fe O}{Ca O''} < \dfrac{3}{5} > \dfrac{1}{7}$.

Dofelic. Dominantly felic, having normative feldspar dominant over normative quartz or lenads. $\dfrac{Q \text{ or } L}{F} < \dfrac{3}{5} > \dfrac{1}{7}$.

Dofemane. Class IV of igneous rocks, having femic minerals dominant over salic. $\dfrac{Sal}{Fem} < \dfrac{3}{5} > \dfrac{1}{7}$.

Dofemic. Dominantly femic, having femic minerals dominant over salic. $\dfrac{Sal}{Fem} < \dfrac{3}{5} > \dfrac{1}{7}$.

Doferrous. Dominantly ferrous, having Fe O dominant over Mg O. $\dfrac{Mg O}{Fe O} < \dfrac{3}{5} > \dfrac{1}{7}$.

Domagnesic. Dominantly magnesic, having Mg O dominant over Fe O. $\dfrac{Mg O}{Fe O} < \dfrac{7}{1} > \dfrac{5}{3}$.

Domalkalic. Dominantly alkalic; of salic minerals when $K_2 O' + Na_2 O'$ dominates over Ca O'. $\dfrac{K_2 O' + Na_2 O'}{Ca O'} < \dfrac{7}{1} > \dfrac{5}{3}$. Of femic minerals when $K_2 O'' + Na_2 O''$ dominates over Mg O + Fe O + Ca O''.

$$\dfrac{Mg O + Fe O + Ca O''}{K_2 O'' + Na_2 O''} < \dfrac{3}{5} > \dfrac{1}{7}$$

202

Domiric. Dominantly miric, having Mg O + Fe O dominant over Ca O''.

$$\frac{Mg\,O + Fe\,O}{Ca\,O''} < \frac{7}{1} > \frac{5}{3}.$$

Domirlic. Dominantly mirlic, having Mg O + Fe O + Ca O'' dominant over K$_2$ O'' + Na$_2$ O''. $\dfrac{Mg\,O + Fe\,O + Ca\,O''}{K_2\,O + Na_2\,O''} < \dfrac{7}{1} > \dfrac{5}{3}.$

Domitic. Dominantly mitic, having mitic minerals (magnetite, hematite, ilmenite, titanite, etc.) dominant over polic minerals (pyroxene, olivine, akermanite).

$$\frac{P\,O}{M} < \frac{3}{5} > \frac{1}{7}.$$

Domolic. Dominantly olic, having normative olivine and akermanite dominant over normative pyroxenes. $\dfrac{P}{O} < \dfrac{3}{5} > \dfrac{1}{7}.$

Dopolic. Dominantly polic, having polic minerals (pyroxene, olivine) dominant over mitic minerals (magnetite, ilmenite, etc.). $\dfrac{P\,O}{M} < \dfrac{7}{1} > \dfrac{5}{3}.$

Dopotassic. Dominantly potassic, having K$_2$ O dominant over Na$_2$ O.

$$\frac{K_2\,O}{Na_2\,O} < \frac{7}{1} > \frac{5}{3}.$$

Dopyric. Dominantly pyric, having normative pyroxene dominant over normative olivine and akermanite. $\dfrac{P}{O} < \dfrac{7}{1} > \dfrac{5}{3}.$

Doquaric. Dominantly quaric, having normative quartz dominant over normative feldspar. $\dfrac{Q}{F} < \dfrac{7}{1} > \dfrac{5}{3}.$

Dosalic. Dominantly salic, having the salic minerals dominant over the femic.

$$\frac{Sal}{Fem} < \frac{7}{1} > \frac{5}{3}.$$

Dosodic. Dominantly sodic, having Na$_2$ O dominant over K$_2$ O.

$$\frac{K_2\,O}{Na_2\,O} < \frac{3}{5} > \frac{1}{7}.$$

E

Extreme. Said of a factor that is present alone or in amount greater than 7:1 of the other factor.

F

Felic. Having the properties of, or containing, the feldspars.

Fem. Term mnemonic of the second group of standard minerals, including non-aluminous ferromagnesian and calcic silicates, silicotitanates and non-siliceous and non-aluminous minerals.

Femic. Having the character of, or belonging to, the second (fem) group of standard minerals.

L

Len. Syllable mnemonic of leucite and nephelite, including sodalite and noselite, the feldspathoids.

Lenad. One of the standard minerals, leucite, nephelite, sodalite or noselite.

M

Magnesiferrous. Equally magnesic and ferrous, or nearly so.

$$\frac{Mg\ O}{Fe\ O} < \frac{5}{3} > \frac{3}{5}.$$

Mir. Syllable mnemonic of magnesia and ferrous iron.

Miric. Characterized by presence of Mg O and Fe O.

Mirl. Syllable mnemonic of magnesia, ferrous iron, and lime.

Mirlic. Characterized by presence of Mg O, Fe O, and Ca O.

Mit. Syllable mnemonic of magnetite, ilmenite, and titanite, and including all minerals of the second subgroup of femic minerals.

Mitic. Adjective referring to the above mentioned minerals.

Mode. The actual mineral composition of a rock. Opposed to norm, with which it may or may not coincide.

O

Ol. Syllable mnemonic of olivine, embracing also akermanite.

Olic. Having the proportions of, or containing, normative olivine or akermanite.

Order. A division of Subclass based on the relative proportions of the standard mineral subgroups in the preponderant group.

P

Per- Prefix to indicate that a factor is present alone, or in extreme amount; that is, its ratio to another factor is $> \dfrac{7}{1}.$

Peralkalic. Extremely alkalic. Of salic minerals when $K_2 O' + Na_2 O'$ is more than seven times Ca O'. $\dfrac{K_2 O' + Na_2 O'}{Ca\ O'} > \dfrac{7}{1}.$ Of femic minerals when $K_2 O'' + Na_2 O''$ is more than seven times Mg O + Fe O + Ca O''.

$$\frac{Mg\ O + Fe\ O + Ca\ O''}{K_2 O'' + Na_2 O''} < \frac{1}{7}.$$

Percalcic. Extremely calcic. Of salic minerals when Ca O' is more than seven times $K_2 O' + Na_2 O'.$ $\dfrac{K_2 O' + Na_2 O'}{Ca\ O'} < \dfrac{1}{7}.$ Of femic minerals when Ca O'' is more than seven times Mg O + Fe O. $\dfrac{Mg\ O + Fe\ O}{Ca\ O''} < \dfrac{1}{7}.$

Perfelic. Extremely felic. When normative feldspar is more than seven times the normative quartz or lenads. $\dfrac{Q\ or\ L}{F} < \dfrac{1}{7}.$

Perfemane. Class V of igneous rocks, having femic minerals extremely abundant.

$$\frac{Sal}{Fem} < \frac{1}{7}.$$

Perfemic. Extremely femic. Having femic minerals more than seven times the salic.

$$\frac{Sal}{Fem} < \frac{1}{7}.$$

Perferrous. Extremely ferrous. When Fe O is more than seven times Mg O.

$$\frac{Mg\ O}{Fe\ O} < \frac{1}{7}.$$

Permagnesic. Extremely magnesic; having Mg O more than seven times Fe O.

$$\frac{Mg\ O}{Fe\ O} > \frac{7}{1}.$$

Permiric. Extremely miric; having Mg O + Fe O more than seven times Ca O''.

$$\frac{Mg\ O + Fe\ O}{Ca\ O''} > \frac{7}{1}.$$

Permirlic. Extremely mirlic; having Mg O + Fe O + Ca O'' more than seven times K$_2$ O'' + Na$_2$ O''. $\quad\dfrac{Mg\ O + Fe\ O + Ca\ O''}{K_2\ O'' + Na_2\ O''} > \dfrac{7}{1}.$

Perolic. Extremely olic; having olic minerals (olivine, akermanite) more than seven times the pyric minerals (pyroxenes). $\dfrac{P}{O} < \dfrac{1}{7}.$

Perpolic. Extremely polic, having polic minerals (pyroxenes, olivine, akermanite) more than seven times the mitic minerals (magnetite, ilmenite, titanite, hematite, etc.). $\quad\dfrac{PO}{M} > \dfrac{7}{1}.$

Perpotassic. Extremely potassic, having K$_2$ O' more than seven times Na$_2$ O'.

$$\frac{K_2\ O'}{Na_2\ O'} > \frac{7}{1}.$$

Perpyric. Extremely pyric, having pyric minerals (pyroxenes) more than seven times the olic minerals (olivine, akermanite). $\dfrac{P}{O} > \dfrac{7}{1}.$

Perquarfelic. Extremely quarfellenic; having normative quartz, feldspar, and feldspathoids more than seven times corundum and zircon. $\dfrac{Q\ F\ L}{C\ Z} > \dfrac{7}{1}.$

Perquaric. Extremely quaric; having normative quartz more than seven times the normative feldspar. $\dfrac{Q}{F} > \dfrac{7}{1}.$

Pol. Syllable mnemonic of the femic silicates pyroxenes and olivine, including akermanite.

Polic. Characterized by the presence of the femic silicates.

Polmitic. Having equal or nearly equal amounts of polic and mitic minerals.

$$\frac{P\ O}{M} < \frac{5}{3} > \frac{3}{5}.$$

Pyr. Syllable mnemonic of pyroxenes.

Pyrolic. Having equal, or nearly equal amounts of normative pyroxene and olivine or akermanite. $\dfrac{P}{O} < \dfrac{5}{3} > \dfrac{3}{5}.$

Q

Quar. Syllable mnemonic of quartz.

Quardofelic. Having felic minerals (feldspar) dominant over normative quartz.

$$\frac{Q}{F} < \frac{7}{1} > \frac{5}{3}.$$

Quarfelic. Having equal or nearly equal amounts of normative quartz and feldspars.

$$\frac{Q}{F} < \frac{5}{3} > \frac{3}{5}.$$

R

Rang. (Old form of rank.) Division of Order based on the character of the chemical bases in the preponderant group of standard minerals.

S

Sal. Syllable mnemonic of the silico-aluminous non-ferromagnesian group of standard minerals, including quartz, feldspars, lenads, corundum and zircon.

Salfemane. Class III of igneous rocks; having salic and femic minerals in equal or nearly equal proportions. $\dfrac{Sal}{Fem} < \dfrac{5}{3} > \dfrac{3}{5}$.

Salfemic. Having salic and femic minerals in equal or nearly equal amounts.

$$\frac{Sal}{Fem} < \frac{5}{3} > \frac{3}{5}.$$

Salic. Having the characters of, or belonging to, the first (sal) group of standard minerals.

Section. Subdivision of any of the other taxonomic divisions from Class to Subgrad.

Subrang. Division of Rang, based on the character of the chemical bases in the preponderant mineral subgroup used in forming Rang.

In order to still further indicate the manner in which the calculations upon which the place of each meteorite in the classification is based are made, two examples of such calculations are here given. The first illustrates the calculation of the mineral components which characterize the great majority of the stony meteorites, the analysis chosen for the calculation being one of the Allegan meteorite made by Stokes.

In the second example is shown the manner of adjusting silica among the different minerals after a preliminary calculation has indicated that too little silica is present to form the more highly siliceous ones. The analysis is one of Felix made by Fireman.

EXAMPLE I

ALLEGAN

Proc. Washington Acad. Sci. 1900, 2, 51

	Per Cent.	Mol.	Apat.	Ilm.	Chrom.	Orth.	Alb.	An.	Diop.	Rem'der.	Hyp.	Oliv.
Si O₂	34.95	583	..	..	..	12	66	24	22	459	262	197
Al₂ O₃	2.55	25	..	..	..	2	11	12	..	..	..	..
Cr₂ O₃	.53	3	..	..	3	..	..	..	..	..	..	..
Fe O	8.47	118	..	1	3	..	..	..				
Mn O	.18	3	..	..	..	..	..	..	11	656	262	394
Mg O......	21.99	550	..	..	..	..	..	..				
Ca O......	1.73	30	7	..	..	..	..	12	11	..	..	..
Na₂ O	.66	11	..	..	..	..	11	..	..	..	..	..
K₂ O	.23	2	..	..	..	2	..	..	..	..	..	..
H₂ O	.25	..	..	..	..	..	..	..	..	..	..	..
Ti O₂	.08	1	..	1	..	..	..	..	..	..	..	..
Fe	21.09	..	..	..	..	..	..	..	..	..	..	..
Ni	1.81	..	..	..	..	..	..	..	..	..	..	..
Co	.05	..	..	..	..	..	..	..	..	..	..	..
Cu	.01	..	..	..	..	..	..	..	..	..	..	..
Fe S	5.05	..	..	..	..	..	..	..	..	..	..	..
P₂ O₅	.27	2	2	..	..	..	..	..	..	..	..	..
Sum.......	100.00	..	..	..	..	..	.	..	..	..	..	..

$$x + y = 656 \ (Mg.\ Fe)\ O$$
$$x + \frac{y}{2} = 459 \ Si\ O_2$$
$$x = 262$$
$$y = 394$$

Formula	Mol. Wt.		Norm

Formula	Mol. Wt.		
$K_2 O.Al_2 O_3.6\ Si\ O_2$ ..	2×556	= orthoclase =	1.11
$Na_2 O.Al_2 O_3.6\ Si\ O_2$.	11×524	= albite =	5.76
$Ca\ O.Al_2 O_3.2\ Si\ O_2$..	12×278	= anorthite =	3.34
$Ca\ O.Si\ O_2$	11×116		
$Mg\ O.Si\ O_2$	9×100	= diopside =	2.44
$Fe\ O.Si\ O_2$	2×132		
$Mg\ O.Si\ O_2$	216×100	= hypersthene =	27.67
$Fe\ O.Si\ O_2$	46×132		
$2\ Mg\ O.Si\ O_2$	325×70	= olivine =	29.79
$2\ Fe\ O.Si\ O_2$	69×102		
$Fe\ O.Cr_2 O_3$	3×224	= chromite =	.67
$Fe\ O.\ Ti\ O_2$	1×152	= ilmenite =	.15
$3\ Ca\ O.P_2 O_5$	2×310	= apatite =	.62
$Fe\ S$		= troilite =	5.05
$Fe_n\ Ni_m$		= nickel-iron =	23.06

F 10.21   Sal 10.21.

P 30.11   P+O 59.90

O

M .82

A 28.73

P+O 59.90   Fem 89.45

$$99.66$$

Perfemic	Dosilicic	Perpolic	Pyrolic
$\dfrac{Sal}{Fem} = \dfrac{10.21}{89.45} < \dfrac{1}{7}$,	$\dfrac{POM}{A} = \dfrac{60.72}{28.73} < \dfrac{7}{1} > \dfrac{5}{3}$,	$\dfrac{PO}{M} = \dfrac{59.90}{.82} > \dfrac{7}{1}$,	$\dfrac{P}{O} = \dfrac{30.11}{29.79} < \dfrac{5}{3} > \dfrac{3}{5}$

Permirlic	Permiric	Domagnesic
$\dfrac{Ca\ O + Mg\ O + Fe\ O}{Na_2 O} = \dfrac{690}{11} > \dfrac{7}{1}$,	$\dfrac{Mg\ O + Fe\ O}{Ca\ O} = \dfrac{667}{23} > \dfrac{7}{1}$,	$\dfrac{Mg\ O}{Fe\ O} = \dfrac{550}{117} < \dfrac{7}{1} > \dfrac{5}{3}$

EXAMPLE II
FELIX
Proc. U. S. Nat. Mus. 1901, 24, 197

	Per Cent.	Mol.	Chromite	Leuc.	Nep.	An.	Tentative Diop.	Tentative Oliv.	Deficit	Ak.	Final Diop.	Final Olv.
Si O$_2$	33.57	560	..	4	20	40	154	437	95	57	2	437
Al$_2$ O$_3$	3.24	31	..	I	10	20	..	..	..	..	..	..
Cr$_2$ O$_3$	.80	5	5	..	..	..	..	..	..	..	..	..
Fe O........	26.22	364	5	..	..		..	..	..	..	..	..
Ni O	1.01	13	..	..	..		77	875	..	..	I	874
Mn O	.68	10	..	..	..		..	..	..	..	..	..
Mg O.......	19.74	493	..	..	..	..	..	..	..	..	..	..
Ca O	5.45	97	..	..	..	20	77	..	..	76	I	..
Na$_2$ O	.62	10	..	..	10	..	..	..	..	..	..	..
K$_2$ O........	.14	I	..	I	..	..	..	..	..	..	..	..
H$_2$ O	.16	..	..	..	..	..	..	..	..	..	..	..
Fe..........	2.59	..	..	..	..	..	..	..	..	..	..	..
Ni..........	.36	..	..	..	..	..	..	..	..	..	..	..
Co	.08	..	..	..	..	..	..	..	..	..	..	..
Cu	.01	..	..	..	..	..	..	..	..	..	..	..
Fe S	4.76	..	..	..	..	..	..	..	..	..	..	..
Graphite ...	.36	..	..	..	..	..	..	..	..	..	..	..
Sum	99.79	..	..	..	..	..	..	..	..	..	..	..

$$2x + 3y + \frac{z}{2} = 496 = \text{available Si O}_2$$
$$x + 4y \quad = \quad 77 = \text{molecules of Ca O}$$
$$x + \quad z = 875 = \text{molecules of Mg O + Fe O.}$$

Whence, $x = 1 = $ diopside, $y = 19 = $ akermanite, $z = 874 = $ olivine.

Formula	Mol. Wt.		Norm				
K$_2$ O.Al$_2$ O$_3$.4 Si O$_2$...	1 × 436	= leucite	= .44	} L 3.28 }			
Na$_2$ O.Al$_2$ O$_3$.2 Si O$_2$.	10 × 284	= nephelite	= 2.84	}		Sal	8.84
Ca O.Al$_2$ O$_3$.2 Si O$_2$..	20 × 278	= anorthite	= 5.56	F 5.56 }			
{ Ca O.Si O$_2$	1 × 116 }						
{ Mg O.Si O$_2$	5 × 100 }	= diopside	= .24	P .24 }			
{ Fe O.Si O$_2$	5 × 132 }						
{ 2 Mg O.Si O$_2$ 492	× 70 }	= olivine	= 73.40	}			
{ 2 Fe O.Si O$_2$ 382	× 102 }			O 81.08 }		Fem 90.24	
4 Ca O.3 Si O$_2$	19 × 404	= akermanite	= 7.68	}			
Fe O.Cr$_2$ O$_3$	5 × 224	= chromite	= 1.12	M 1.12			
Fe S.............		= troilite	= 4.76	} A 7.80 }			
Fe$_n$ N$_m$		nickel-iron	= 3.04	}			
		H$_2$ O	= .16				
		graphite	= .36				

$$99.60$$

Perfemic · Persilicic · Perpolic

$$\frac{\text{Sal}}{\text{Fem}} = \frac{8.84}{90.24} < \frac{1}{7}, \quad \frac{\text{POM}}{\text{A}} = \frac{82.44}{7.80} > \frac{7}{1}, \quad \frac{\text{PO}}{\text{M}} = \frac{81.32}{1.12} > \frac{7}{1}$$

Perolic · Permirlic · Permiric

$$\frac{\text{P}}{\text{O}} = \frac{.24}{81.08} < \frac{1}{7}, \quad \frac{\text{Ca O + Mg O + Fe O}}{\text{Na}_2\text{ O}} = \frac{977}{10} > \frac{7}{1}, \quad \frac{\text{Mg O + Fe O}}{\text{Ca O}} = \frac{880}{97} > \frac{7}{1}$$

Magnesiferrous

$$\frac{\text{Mg O}}{\text{Fe O}} = \frac{493}{387} < \frac{5}{3} > \frac{3}{5}$$

ALPHABETICAL LIST OF THE STONE METEORITES ANALYSES OF WHICH ARE GIVEN

The numbers refer to the number of the analysis in the following table of analyses

In some cases different analyses of the same meteorite require it to be placed in more than one group. Such cases indicate that further analyses are needed. In Busti for example there seems to be no way of determining whether Dancer's or Maskelyne's analysis is the more nearly correct and both must be used, but further analyses would probably furnish ground for eliminating one or the other. It is quite possible that a similar confusion would appear in terrestrial rocks if analyses of the same rock made at widely different times and by different analysts were compared. While some such discrepancies occur, in most cases plural analyses agree in placing the meteorite in the same group. This is true for example, of Homestead, New Concord, Aussun, Hessle, and others. In such cases the plurality of analyses happily confirms the placing of the meteorite. An opportunity for comparison of the grouping of meteorites in the quantitative classification with that of Rose, Tschermak, and Brezina is afforded by the Brezina symbol of each meteorite given in the tables. Comparison shows that on the whole the important groups of the German classification remain intact in the quantitative classification. Thus the howarditites, eukrites, and chladnites occupy on the whole similar and separate places in both classifications. Among the subgroups of the chondrites little similarity of grouping in the two classifications can be noted, though the gray chondrites and spherical chondrites are rather more numerous among the less siliceous groups of the quantitative classification. This would be expected since the color and structure of the meteorites of these groups indicate a larger proportion of olivine than in the white or intermediate chondrites. Such a scattering of these groups, however, on the whole emphasizes the impossibility of accurately classifying meteorites by their physical characters as has hitherto been attempted by the German system.

An interesting feature of the calculations is the indication which they afford of the presence of leucite or nephelite or both in some meteorites, such as Felix, Shytal, and Cold Bokkeveld. The calculation of these minerals was required by the low percentage of silica and suggests that a careful examination of the meteorites for these minerals, which have not been hitherto observed in meteorites, should be made. The most common meteorite type is seen from the tables to be that of Pultusk, perfemic, dosilicic, perpolic, pyrolic, permirlic, permiric, and domagnesic.

The Farmington type is also largely represented, differing from Pultusk only in being domolic instead of pyrolic. Further it will be seen by examining the tables that the great majority of meteorites are domagnesic and in making the calculations it was found that a proportion of Mg O to Fe O of very nearly 4:1 was highly preponderant and characteristic.

A summation of all the analyses, 125 in number, should give a fair average of the composition of stone meteorites. It gives the following result:

AVERAGE COMPOSITION OF STONE METEORITES

$Si O_2$	39.12
$Al_2 O_3$	2.62
$Fe_2 O_3$	.38
$Cr_2 O_3$	.41
Fe O	16.13
Mn O	.18
Ni O	.21
Mg O	22.42
Ca O	2.31
$Na_2 O$	.81
$K_2 O$	.20
$H_2 O$	.20
Fe	11.46
Ni	1.15
Co	.05
S	1.98
P	.04
$P_2 O_5$	.03
C	.06
Ni, Mn, Cu, Sn	.02
$Ti O_2$	.02
$Sn O_2$	.02
	99.82

The results agree very nearly with those obtained by Merrill* by the addition of 99 analyses, the principal difference being a larger percentage of Ca O in the present writer's result. The present writer's method of determining the minor constituents differed from that of Merrill in that the present writer divided the totals of these constituents by the total number of analyses instead of by the number of analyses in which each constituent was reported. It is evident that the writer's method will produce too low a result, but the other method may give one too high, since the minor constituents may have been lacking in analyses in which they were not reported. It may further be suggested by way of discussion of the interesting comparison made by Merrill between stony meteorites and the earth's crust, that only the lighter and more siliceous meteorites should be used for such a comparison. Stony meteorites having large percentages of free metal have too high a specific gravity to be strictly comparable with the earth's crust. Again it should be recognized that the greater abundance of certain elements at the surface of the earth may be on account of their greater solubility. Thus limestones have grown successively more calcic and less magnesian since early times and an increase in the amount of soda and potash at the surface might take place in the same way. It does not appear that such a process would explain the discrepancy in the amount of alumina but it might act to increase the amount of silica. That the earth's crust of earlier times was more nearly meteoritic in composition than the present seems to be indicated by the great deposits of iron oxide of earlier ages and the fact that the early limestones are more magnesian than the modern.

Adding the analyses of iron meteorites p. 229 to those previously published, and omitting about 60 obviously imperfect ones, 318 analyses are obtained from which the average composition of iron meteorites can be calculated by summation. This sum is as follows:

AVERAGE COMPOSITION OF IRON METEORITES

Fe.. 90.85
Ni.. 8.52
Co.. .59
P... .17
S... .04
C... .03
Cu.. .02
Cr.. .01
_____
100.23

* Am. Jour. Sci. 1909, 4. 27, 471.

Combining this sum with that previously obtained from 125 analyses of stone meteorites, stone meteorites being here regarded as all those which have an appreciable quantity of silicates, the sum total gives according to Clarke's method* the average composition of meteorites as a whole. The method is, of course, empirical, but seems to be the only one available in our present state of knowledge. This sum is the following:

AVERAGE COMPOSITION OF METEORITES

Fe	68.43
$Si O_2$	11.07
Ni	6.44
$Mg O$	6.33
Fe O	4.55
$Al_2 O_3$	.74
Ca O	.65
S	.49
Co	.44
$Na_2 O$	.23
P	.14
$Cr_2 O_3$	.12
$Fe_2 O_3$	.11
Ni O	.06
$K_2 O$	.05
Mn O	.04
C	.04
Cu	.01
Cr	.01
$P_2 O_5$	.01
$Ti O_2$	.01
$Sn O_2$	.01
	99.98

The present writer has previously suggested,† that the average composition of meteorites may represent the composition of the earth as a whole. If so the proportions of the elements in the earth as a whole would be as follows:

PROPORTION OF ELEMENTS IN THE EARTH AS A WHOLE AS DEDUCED FROM METEORITES

Iron	72.06
Oxygen	10.10
Nickel	6.50
Silicon	5.20

* Bull. U. S. Geol. Survey, 1891, 78, 33.
† Jour. Geol. 1901, 9, 630.

Magnesium	3.80
Sulphur	.49
Calcium	.46
Cobalt	.44
Aluminum	.39
Sodium	.17
Phosphorus	.14
Chromium	.09
Potassium	.04
Carbon	.04
Manganese	.03
Other elements	.05
	100.00

The large proportion of iron in the constitution of the earth indicated by meteorites is in accord with the earth's density, rigidity, and magnetic proportions. Assuming the density of the rocks of the earth's crust to be 2.8, which may be too high, and combining with it metal of the density of 7.8, which is an average of the density of iron meteorites, it will be found that 77.58 per cent of metal will be required to obtain a density of 5.57, that of the earth as a whole. This is very nearly that of the sum of the metals in the above result after eliminating the proportions present as oxides. Such a proportion of iron would seem to be in accord, as has been stated, with the earth's rigidity and magnetic properties.

SYNOPSIS OF METEORITE CLASSIFICATION

CLASS III. $\dfrac{Sal}{Fem} < \dfrac{5}{3} > \dfrac{3}{5}$

SALFEMIC

SUBCLASS I. $\dfrac{QFL}{CZ} > \dfrac{7}{1}$

PERQUARFELIC

Order	1. $\dfrac{Q}{F} > \dfrac{7}{1}$ Perquaric	2. $\dfrac{Q}{F} < \dfrac{7}{1} > \dfrac{5}{3}$ Doquaric	3. $\dfrac{Q}{F} < \dfrac{5}{3} > \dfrac{3}{5}$ Quarfelic	4. $\dfrac{Q}{F} < \dfrac{3}{5} > \dfrac{1}{7}$ Quardofelic	5. $\dfrac{QL}{F} < \dfrac{1}{7}$ Perfelic
Rang 1. Peralkalic, $\dfrac{K_2O + Na_2O}{CaO} > \dfrac{7}{1}$					
Rang 2. Domalkalic, $\dfrac{K_2O + Na_2O}{CaO} < \dfrac{7}{1} > \dfrac{5}{3}$					
Rang 3. Alkalicalcic, $\dfrac{K_2O + Na_2O}{CaO} < \dfrac{5}{3} > \dfrac{3}{5}$					
Rang 4. Docalcic, $\dfrac{K_2O + Na_2O}{CaO} < \dfrac{3}{5} > \dfrac{1}{7}$					
Rang 5. Percalcic, $\dfrac{K_2O + Na_2O}{CaO} < \dfrac{1}{7}$					Juvinose

CLASS IV. $\dfrac{Sal}{Fem} < \dfrac{3}{5} > \dfrac{1}{7}$

DOFEMIC

Section	SUBCLASS I. $\frac{POM}{A} > \frac{7}{1}$ PERSILICIC — ORDER 1. $\frac{P.O}{M} > \frac{7}{1}$ PERFELIC					SUBCLASS II. $\frac{POM}{A} < \frac{7}{1} > \frac{5}{3}$ POSILICIC — ORDER 1. $\frac{P.O}{M} > \frac{7}{1}$ PERFELIC					ORDER 2. $\frac{P.O}{M} < \frac{7}{1} > \frac{5}{3}$ DOFELIC			SUBCLASS III. $\frac{POM}{A} < \frac{7}{1} > \frac{5}{3}$ SILICOMETALLIC — ORDER 1. $\frac{P.O}{M} > \frac{7}{1}$ PERFELIC				
	1. $\frac{P}{O}>\frac{7}{1}$ Perpyric	2. $\frac{P}{O}<\frac{7}{1}>\frac{5}{3}$ Dopyric	3. $\frac{P}{O}<\frac{5}{3}>\frac{3}{5}$ Pyrolic	4. $\frac{P}{O}<\frac{3}{5}>\frac{1}{7}$ Domolic	5. $\frac{P}{O}<\frac{1}{7}$ Perolic	1. $\frac{P}{O}>\frac{7}{1}$ Perpyric	2. $\frac{P}{O}<\frac{7}{1}>\frac{5}{3}$ Dopyric	3. $\frac{P}{O}<\frac{5}{3}>\frac{3}{5}$ Pyrolic	4. $\frac{P}{O}<\frac{3}{5}>\frac{1}{7}$ Domolic	5. $\frac{P}{O}<\frac{1}{7}$ Perolic	1. $\frac{P}{O}>\frac{7}{1}$ Perpyric	2. $\frac{P}{O}<\frac{7}{1}>\frac{5}{3}$ Dopyric	3, 4 and 5 not represented	1. $\frac{P}{O}>\frac{7}{1}$ Perpyric	2. $\frac{P}{O}<\frac{7}{1}>\frac{5}{3}$ Dopyric	3. $\frac{P}{O}<\frac{5}{3}>\frac{3}{5}$ Pyrolic	4. $\frac{P}{O}<\frac{3}{5}>\frac{1}{7}$ Domolic	5. $\frac{P}{O}<\frac{1}{7}$ Perolic
Rang I. Permitic. $\frac{CaO+MgO+FeO}{Na_2O}>\frac{7}{1}$																		
Section 1. Permitic. $\frac{MgO+FeO}{CaO}>\frac{7}{1}$																		
Subrang 1. Permagnesic, $\frac{MgO}{FeO}>\frac{7}{1}$																		
Subrang 2. Domagnesic, $\frac{MgO}{FeO}<\frac{7}{1}>\frac{5}{3}$		Sherpotose	Udenose				Linumose									Borkutose		
Subrang 3. Magnesiferrous, $\frac{MgO}{FeO}<\frac{5}{3}>\frac{3}{5}$					Stawropolose		Krahmbergose	Parnallose	Estacadose	Albartiose		Pickerase				Income	Kernouvose	
Subrang 4. Doferrous, $\frac{MgO}{FeO}<\frac{3}{5}>\frac{1}{7}$																		
Subrang 5. Perferrous, $\frac{MgO}{FeO}<\frac{1}{7}$																		
Section 2. Domitic, $\frac{MgO+FeO}{CaO}<\frac{7}{1}>\frac{5}{3}$	Frankfortose																	
Subrang 1. Permagnesic, $\frac{MgO}{FeO}>\frac{7}{1}$	Stauzernose																	
Subrang 2. Domagnesic, $\frac{MgO}{FeO}<\frac{7}{1}>\frac{5}{3}$																		
Subrang 3. Magnesiferrous, $\frac{MgO}{FeO}<\frac{5}{3}>\frac{3}{5}$																		
Section 3. Calcimitic, $\frac{MgO+FeO}{CaO}<\frac{5}{3}>\frac{1}{7}$			Angrose															
Subrang 1. Permagnesic, $\frac{MgO}{FeO}>\frac{7}{1}$																		
Subrang 2. Domagnesic, $\frac{MgO}{FeO}<\frac{7}{1}>\frac{5}{3}$																		
Subrang 3. Magnesiferrous, $\frac{MgO}{FeO}<\frac{5}{3}>\frac{3}{5}$																		
Subrang 4. Doferrous, $\frac{MgO}{FeO}<\frac{3}{5}>\frac{1}{7}$	Constantinophose																	

216

CLASS V. $\frac{Sal}{Fem} < \frac{1}{7}$
PERSALIC

SUBCLASS I. $\frac{POM}{A} > \frac{7}{1}$ — PERSALIC — ORDER 1. $\frac{PO}{M} > \frac{7}{1}$ PERPOLIC

Section	1. $\frac{P}{O} > \frac{7}{1}$ Perpyric	2. $\frac{P}{O} < \frac{7}{1} > \frac{5}{3}$ Dopyric	3. $\frac{P}{O} < \frac{5}{3} > \frac{3}{5}$ Pyrolic	4. $\frac{P}{O} < \frac{3}{5} > \frac{1}{7}$ Domolic	5. $\frac{P}{O} < \frac{1}{7}$ Perolic
Rang 1. Persalic, $\frac{CaO+MgO+FeO}{Na_2O} > \frac{7}{1}$					
Section 1. Persalic, $\frac{MgO+FeO}{CaO} > \frac{7}{1}$					
Subrang 1. Permagnesic, $\frac{MgO}{FeO} > \frac{7}{1}$	Bishopvillose	×			
Subrang 2. Domagnesic, $\frac{MgO}{FeO} < \frac{7}{1} > \frac{5}{3}$	Thhembehirranose	Shallose	Travisose	Wacondose	Kakovose
Subrang 3. Magnesiferrous, $\frac{MgO}{FeO} < \frac{5}{3} > \frac{3}{5}$		Middlesborose	Concordose	Kabose	Jeromose
Subrang 4. Doferrous, $\frac{MgO}{FeO} < \frac{3}{5} > \frac{1}{7}$					
Subrang 5. Perferrous, $\frac{MgO}{FeO} < \frac{1}{7}$					
Section 2. Domic, $\frac{MgO+FeO}{CaO} < \frac{7}{1} > \frac{5}{3}$		Bentose			
Subrang 1. Permagnesic, $\frac{MgO}{FeO} > \frac{7}{1}$					
Subrang 2. Domagnesic, $\frac{MgO}{FeO} < \frac{7}{1} > \frac{5}{3}$					
Subrang 3. Magnesiferrous, $\frac{MgO}{FeO} < \frac{5}{3} > \frac{3}{5}$					

ORDER 2. $\frac{PO}{M} < \frac{7}{1} > \frac{5}{3}$ DOPOLIC

Section	1. $\frac{P}{O} > \frac{7}{1}$ Perpyric	2. $\frac{P}{O} < \frac{7}{1} > \frac{5}{3}$ Dopyric	3. $\frac{P}{O} < \frac{5}{3} > \frac{3}{5}$ Pyrolic
			Elwahose

SUBCLASS II. $\frac{POM}{A} < \frac{7}{1} > \frac{5}{3}$ — DOSALIC — ORDER 1. $\frac{PO}{M} > \frac{7}{1}$ PERPOLIC

Section	1. $\frac{P}{O} > \frac{7}{1}$ Perpyric	2. $\frac{P}{O} < \frac{7}{1} > \frac{5}{3}$ Dopyric	3. $\frac{P}{O} < \frac{5}{3} > \frac{3}{5}$ Pyrolic	4. $\frac{P}{O} < \frac{3}{5} > \frac{1}{7}$ Domolic	5. $\frac{P}{O} < \frac{1}{7}$ Perolic
	Hvittisose		Orviniose		
	Mocsose	Castaliose	Pultuskose	Farmingtonose	Ornansose
		Ensisheimose	Homesteadose		

SUBCLASS III. $\frac{POM}{A} < \frac{5}{3} > \frac{3}{5}$ — SILICOMETALLIC

Not represented

SUBCLASS IV. $\frac{POM}{A} < \frac{3}{5} > \frac{1}{7}$ — DOMETALLIC — ORDER 1. $\frac{PO}{M} > \frac{7}{1}$ PERPOLIC

Section	1. $\frac{P}{O} > \frac{7}{1}$ Perpyric	2. $\frac{P}{O} < \frac{7}{1} > \frac{5}{3}$ Dopyric	3 and 4 not represented	5. $\frac{P}{O} < \frac{1}{7}$ Perolic
	Steinbichose	Minciose		Marjalahtose

ANALYSES OF STONE METEORITES

COMPILED AND CLASSIFIED ACCORDING TO THE PRINCIPLES OF THE AMERICAN QUANTITATIVE CLASSIFICATION

CLASS III

SALFEMIC, PERQUARFELIC, PERFELLIC, PERCALCIC, JUVINOSE

Name	SiO₂	Al₂O₃	FeO	MgO	CaO	Na₂O	K₂O	Fe	Ni	Co	S	P	Miscellaneous	Sum	Sp. gr.	Norm	Brezina's Symbol	Analyst	Reference
1. Juvinas	49.23	12.55	20.33	6.44	10.23	0.63	0.12	0.16	....	....	0.09	....	FeS 0.1 21 TiO₂ 0.10 Cr₂O₃ 0.24 P₂O₅ 0.28	101.61	3.12	Q 2.2 di 14.4 cm 1.6 / or 0.6 hy 44.2 tr 0.6 / ab 5.2 ml 1.9 / an 31.1	Eu	C. Rammelsberg	Ann. Phys. Chem. 1848, 77, 585–590

CLASS IV

DOFEMIC, PERSILICIC, PERPOLIC, PERPYRIC, PIRMIRLIC, DOMIRIC, DOMAGNESIC, FRANKFORTOSE

Name	SiO₂	Al₂O₃	FeO	MgO	CaO	Na₂O	K₂O	Fe	Ni	Co	S	P	Miscellaneous	Sum	Sp. gr.	Norm	Brezina's Symbol	Analyst	Reference
2. Frankfort	51.33	8.05	13.70	17.59	7.03	0.45	0.22	tr	tr	....	0.23	....	Cr₂O₃ 0.42	99.02	3.31	or 1.1 di 12.4 cm 0.7 / ab 3.7 hy 59.4 tr 0.6 / an 19.5 ol 6.3	Ho	G. J. Brush and W. J. Mixter	Am. Jour. Sci. 1869, 2, 48, 243

DOFEMIC, PERSILICIC, FERPOLIC, PERPYRIC, PERIRLIC, DOMIRIC, MAGNESIFERROUS, STANNERNOSE

Name	SiO₂	Al₂O₃	FeO	MgO	CaO	Na₂O	K₂O	Fe	Ni	Co	S	P	Miscellaneous	Sum	Sp. gr.	Norm	Brezina's Symbol	Analyst	Reference
3. Mässing	53.12	8.20	19.14	8.48	5.79	1.93	1.19	0.52	....	....	0.37	....	Cr₂O₃ 0.08	99.72	3.36	Q 1.3 di 15.9 cm 1.6 / or 7.2 hy 45.8 tr 0.6 / ab 10.2 ml 0.6 / an 10.0	Ho	A. Schwager	Sitzber. München Akad. 1878, 8, 32–40
4. Petersburg	49.21	11.05	20.41	8.13	9.01	0.82	....	0.50	tr	....	0.06	....		99.23	3.20	Q 0.1 di 15.5 tr 0.2 / ab 6.8 hy 49.5 ml 0.5 / an 26.4	Ho	J. L. Smith	Am. Jour. Sci. 1861, 2, 31, 265
5. Peramiho	49.32	11.24	20.65	7.15	10.84	0.40	0.25	....	....	....	0.23	....	TiO₂ 0.42	100.50		Q 1.2 di 21.7 il 0.8 / ab 3.1 hy 43.3 tr 0.6 / an 28.1	Eu	E. Ludwig	Sitzber. Wien Akad. 1903, 112, 739–777
6. Stannern	48.30	12.65	19.32	6.87	11.27	0.62	0.23	....	....	....	tr	....	Chromite 0.54 MnO 0.81	100.61	3.05	or 1.1 d 20.0 cm 0.5 / ab 5.2 hy 35.0 / an 31.1 ol 5.6	Eu	C. Rammelsberg	Ann. Phys. Chem. 1851, 83, 591–593

DOFEMIC, PERSILICIC, PERPOLIC, PERPYRIC, PERMRLIC, CALCIMIRIC, DOPERROUS, CONSTANTINOPLOSE

Name	SiO₂	Al₂O₃	FeO	MgO	CaO	Na₂O	K₂O	Fe	Ni	Co	S	P	Miscellaneous	Sum	Sp. gr.	Norm	Brezina's Symbol	Analyst	Reference
7. Constantinople	48.59	12.63	20.99	6.16	10.39	0.46	0.16	....	....	....	....	....	Cr₂O₃ 0.44 Mn O tr	99.82		Q 0.5 di 16.8 / or 1.7 hy 45.3 / ab 3.7 cm 0.4 / an 32.0	Eu	G. Tschermak	Min. Mitth. 1872, 2, 85

DOFEMIC, PERSILICIC, PERPOLIC, DOPYRIC, PERIRLIC, DOMIRIC, MAGNESIFERROUS, SHERGOTTOSE

Name	SiO₂	Al₂O₃	FeO	MgO	CaO	Na₂O	K₂O	Fe	Ni	Co	S	P	Miscellaneous	Sum	Sp. gr.	Norm	Brezina's Symbol	Analyst	Reference
8. Shergotty	50.21	5.90	21.85	10.00	10.41	1.28	0.57	....						100.22		or 3.3 di 26.2 / ab 11.0 hy 24.5 / an 8.6 ol 16.7	She	E. Lumpe	Min. Mitth. 1871, 55–56

ANALYSES OF STONE METEORITES -- *Continued*

DOFEMIC, PERSILICIC, PERPOLIC, PYROLIC, PERMIRIC, PERMIRLIC, PERMIRIC, DOMAGNESIC, UDENOSE

Name	SiO₂	Al₂O₃	FeO	MgO	CaO	Na₂O	K₂O	Fe	Ni	Co	S	P	Miscellaneous	Sum	Sp.gr.	Norm	Brezina's Symbol	Analyst	Reference
9. Uden	44.58	4.10	22.41	20.67	2.28	0.94	0.49	1.77			Fe S 0.72		Chromite 0.76 MnO 0.43 NiO 0.39	99.44	3.40	or 2.8 di 4.8 mi 0.8 / ab 7.0 hy 29.7 tr 0.7 / an 5.6 ol 45.4 nf 1.8	Cwb	Baumhauer and Seelheim	Ann. Phys. Chem. 1862, 116, 185–188
10. Knyahinya	44.30	3.06	16.38	22.16	2.73	1.00	0.66	5.00			Fe S 2.22		Chromite 0.80	98.31	3.52	or 3.0 di 0.5 cm 0.8 / ab 8.4 hy 28.0 tr 2.2 / an 2.0 ol 37.6 nf 5.0	Cg	E. H. von Baumhauer	Arch. Neerland, 1872, 7, 146–153. Mass anal. calc. by Wadsworth

DOFEMIC, PERSILICIC, PERPOLIC, PYROLIC, PERMIRLIC, CALCIMIRIC, DOMAGNESIC, ANGROSE

Name	SiO₂	Al₂O₃	FeO	MgO	CaO	Na₂O	K₂O	Fe	Ni	Co	S	P	Miscellaneous	Sum	Sp.gr.	Norm	Brezina's Symbol	Analyst	Reference
11. Angra dos Reis	43.94	8.73	8.28	10.05	24.51	0.26	0.19	0.81			0.45		Fe O 0.31 Ti O 1.30 Pr O 0.13	100.05	...	le 0.0 di 35.1 mi 0.5 / ne 1.1 ol 15.7 ap 0.3 / an 22.0 cm 20.2 nf 0.8	Angrite	Ludwig and Tschermak	Min. u. petr. Mitth. N. F. 1909, 28, 113

DOFEMIC, PERSILICIC, PERPOLIC, PEROLIC, PERMIRIC, PERMIRIC, DOMAGNESIC, STAWROPOLOSE

Name	SiO₂	Al₂O₃	FeO	MgO	CaO	Na₂O	K₂O	Fe	Ni	Co	S	P	Miscellaneous	Sum	Sp.gr.	Norm	Brezina's Symbol	Analyst	Reference
12. Stawropol	33.16	4.22	18.59	29.24	1.20	1.40	0.60	4.32			1.60		NiO 3.81 SnO 1.10	99.24	3.59	le 2.6 ol 71.0 tr 4.4 / net 6.5 cm 0.8 nf 4.3 / an 3.3 mo 4.3	Ck	H. Abich	Bull. Akad. St. Petersburg, 1860, 1862, 403–422, 433–430

DOFEMIC, DOSILICIC, PERPOLIC, DOPYRIC, PERMIRIC, PERMAGNESIC, LINUMOSE

Name	SiO₂	Al₂O₃	FeO	MgO	CaO	Na₂O	K₂O	Fe	Ni	Co	S	P	Miscellaneous	Sum	Sp.gr.	Norm	Brezina's Symbol	Analyst	Reference
13. Linum	43.05	2.44	1.32	25.72	3.49	1.39	0.26	15.83	0.71		1.85	0.07	Cr₂O₃ 0.31 MnO 0.20 H₂O 0.12 Fe S 3.23	99.99	3.54	or 7.6 ns 0.2 cm 0.5 / ab 11.0 ol 5.0 tr 1.5 / hy 43.0 th 3.0 ol 13.3 nf 16.55	Cw	Lindner	Sitzber. Berlin Akad. 1904, 114–153

DOFEMIC, DOSILICIC, PERPOLIC, DOPYRIC, PERMIRIC, PERMIRIC, DOMAGNESIC, KRÄHENBERGOSE

Name	SiO₂	Al₂O₃	FeO	MgO	CaO	Na₂O	K₂O	Fe	Ni	Co	S	P	Miscellaneous	Sum	Sp.gr.	Norm	Brezina's Symbol	Analyst	Reference
14. Krähenberg	41.12	3.22	17.42	18.62	2.06	0.17	1.22	10.37	1.36		2.35	0.46	Cr₂O₃ 0.80 MnO 0.78 SnO 0.18	100.22	...	or 7.2 di 4.9 cm 1.3 / ab 1.6 hy 44.0 tr 0.4 / an 4.2 ol 16.2 nf 11.7	Cho	Keller	Sitzber. München Akad. 1878, 8, 47–58

DOFEMIC, DOSILICIC, PERPOLIC, PYROLIC, PERMIRIC, PERMIRIC, DOMAGNESIC, PARNALLOSE

Name	SiO₂	Al₂O₃	FeO	MgO	CaO	Na₂O	K₂O	Fe	Ni	Co	S	P	Miscellaneous	Sum	Sp.gr.	Norm	Brezina's Symbol	Analyst	Reference
15. Lesves	39.46	3.33	15.82	22.75	1.54	1.05	0.09	12.36	1.37	0.11	2.25		Cr₂O₃ 1.02	101.15	3.58	or 0.6 di 3.1 cm 1.6 / ab 8.9 hy 39.0 tr 6.2 / an 3.9 ol 31.0 nf 13.8	Cw	A. F. Renard	Bull. de l'Acad. roy. de Belgique, 1896, 3, 31, 654–663
16. Parnallee	39.41	2.57	15.28	22.82	0.56	0.55		9.83	0.90	0.06	2.71	0.10	MnO 0.54 FeO 0.68 NiO 0.72 CoO 0.06	98.70	3.12	or 2.8 ns 1.3 tr 7.4 / ab di 4.2 nf 10.8 / an 35.5	Cga	E. Pfeiffer	Sitzber. Wien. Akad. 1863, 47, 2, 460–463
17. Carcote	39.28	2.39	14.29	22.79	1.19	1.40	0.30	8.95	0.91		Fe S 5.98	0.21	Chromite 1.43 Cu+Sn 0.06 MnO 0.14 Co 0.10 Res 0.49	100.00	3.47	or 1.7 di 0.2 cm 1.4 / ab 11.0 hy 4.7 tr 6.0 / an 22.3 ol 40.4 nf 10.1	Ck	Will	Neues Jahrb. 1889, 2, 177–179. Mass anal. calc. by Farrington

219

DOFEMIC, DOSILICIC, PERPOLIC, DOMOLIC, PERMIRLIC, PERMIRIC, DOMAGNESIC, ESTACADOSE

Name	SiO₂	Al₂O₃	FeO	MgO	CaO	Na₂O	K₂O	Fe	Ni	Co	S	P	Miscellaneous	Sum	Sp. gr.	Norm	Brezina's Symbol	Analyst	Reference
18. Bjurböle	41.06	2.55	13.80	25.75	1.82	1.24	0.32	6.38	0.72	0.04	Fe S 5.44	0.14	Cr₂O₃ 0.59, MnO 0.17, NiO 0.07	100.04	...	or 1.7 di 6.3 cm 0.9 / ab 10.5 hy 18.4 tr 5.4 / ol 47.8 nf 7.1	Cca	Ramsay and Borgström	Bull. Com. Geol. de Finland, 1902, 12, 13
19. Nerft	40.00	3.52	15.98	25.59	0.05	1.65	0.08	8.36	1.32	tr	2.02	0.05	Chromite 0.65, MnO 0.03, MnO 0.10	99.40	...	or 0.6 hy 21.1 cm 0.7 / ab 14.1 ol 45.2 tr 5.5 / an 0.3 nf 9.8 / C 0.5	Cia	A. Kuhlberg	Ann. Phys. Chem. 1869 136, 448-449
20. Rakowka	38.87	2.66	13.44	24.60	2.36	2.04	0.37	5.67	1.43	0.32	Fe S 6.16		C 0.13, Mn tr	99.22	3.58	or 2.5 mt 1.3 cm 0.8 / ab 11.5 di 5.2 sc 0.8 / hy 6.2 tr 6.2 / ol 54.0 nf 7.4	Ci	P. Grigorieux	Zeitschr. deutsch. Geol. Gesell. 1880, 32, 417-420
21. Chandakapur	38.02	4.17	19.81	21.31	2.42	1.26	0.29	5.25	0.55		Fe S Fe₃P 4.92	Fe₃P 1.06	Chromite 0.51, NiO 0.07	99.94	...	or 1.7 di 5.7 mt 0.3 / ab 10.5 an 5.0 cm 0.5 / ol 60.5 tr 4.9 / nf 5.8	Cib	H. E. Clarke	Min. Mag. 1910, 15, 371
22. Mező-Madaras	37.64	3.41	15.44	24.11	1.68	1.76	tr	12.12	1.64		2.27		Chromite 0.54, MnO 0.8, NiO 0.06	100.85	...	ab 14.8 di 5.6 cm 0.7 / an 1.4 hy 8.1 tr 6.3 / ol 49.1 nf 13.8	Cgb	C. Rammelsberg	Zeitschr. deutsch. Geol. Gesell. 1871, 23, 734-737. Mass anal. calc. by Wadsworth
23. Tourinnes-la-Grosse	37.47	3.65	13.89	24.40	2.61	2.26		11.05	1.30		2.21		Chromite 0.71, Sn 0.17	99.72	3.53	ab 15.2 ms 0.1 cm 0.7 / ne 2.0 di 10.2 tr 6.1 / ol 51.9 nf 12.5	Cw	F. Pisani	Comptes Rendus, 1864, 58, 169-171
24. Meuselbach	37.30	2.89	16.20	24.55	1.72	1.32		6.71	1.07	0.11	Fe S 7.79		Chromite 0.34, Cn tr	100.00	...	ab 11.0 di 5.2 cm 0.3 / an 2.0 ol 59.7 tr 7.8 / nf 7.9	Ccka	G. Linck	Ann. Wien. Mus. 1899, 13, 103-114, Mass anal. calc. by Farrington
25. Lundsgård	36.97	2.70	13.18	23.79	1.40	1.42	0.43	14.46	1.91	0.02	2.38	0.10	Chromite 0.59, NiO 0.05, H₂O 0.30, C 0.02	99.96	3.61	or 2.2 ms 0.1 cm 0.9 / ab 11.5 hy 15.8 tr 6.5 sc 0.6 / ol 38.4 nf 16.4	Cw	O. Nordenskjöld	Geol. Foren. i. Stockholm, Förh. 1891, 13, 470-475
26. Estacado	35.82	3.60	15.53	22.74	2.99	2.07	0.32	14.68	1.60	0.08	1.37	0.15	Cr₂O₃ tr, MnO tr, TiO₂ tr, Cu tr	100.95	3.60	or 1.7 tr 3.8 / ab 0.4 di 0.1 sc 1.0 / hy 12.0 nf 16.4 / ol 51.5	Ckb	J. M. Davison	Am. Jour. Sci. 1906, 3, 22, 59

DOFEMIC, DOSILICIC, PERPOLIC, PEROLIC, PERMIRLIC, PERMIRIC, PERMIRIC, DOMAGNESIC, ALBARETOSE

Name	SiO₂	Al₂O₃	FeO	MgO	CaO	Na₂O	K₂O	Fe	Ni	Co	S	P	Miscellaneous	Sum	Sp. gr.	Norm	Brezina's Symbol	Analyst	Reference
27. Albareto	35.91	4.48	24.31	22.77	2.07	1.64	0.44	4.33	0.73	0.11	2.37			99.16	...	or 2.2 di 5.2 tr 6.5 / ab 5.8 ol 64.8 nf 5.2 / ne 4.3 / an 4.9	Cc	P. Maissen	Gazetta Chimica, 1880, 10, 20

DOFEMIC, DOSILICIC, DOPOLIC, DOPYRIC, PERMIRLIC, PERMIRIC, PERMAGNESIC, PICKENSOSE

Name	SiO₂	Al₂O₃	FeO	MgO	CaO	Na₂O	K₂O	Fe	Ni	Co	S	P	Miscellaneous	Sum	Sp. gr.	Norm	Brezina's Symbol	Analyst	Reference
28. Pickens County	37.06	5.83	9.63	24.00	0.55	0.92	0.02	8.22	1.23	0.11	1.57		Fe₂O₃ 0.69, MnO 0.49, CrO 0.36, CuO 0.06, TiO₂ 0.09, P₂O₅ 0.31	101.05	...	ab 7.9 ky 42.0 ms 15.5 / an 0.8 il 15.2 il 0.2 / C 4.0 tr 0.7 / tr 4.3 / nf 9.6		E. Everhart	Science, 1909, N. S. 30, 772

DOFEMIC, SILICOMETALLIC, PERPOLIC, PYROLIC, PERMIRLIC, PERMIRIC, PERMAGNESIC, BORKUTOSE

Name	SiO₂	Al₂O₃	FeO	MgO	CaO	Na₂O	K₂O	Fe	Ni	Co	S	P	Miscellaneous	Sum	Sp. gr.	Norm	Brezina's Symbol	Analyst	Reference
29. Borkut	35.28	2.74	4.71	19.92	1.95	1.91	0.66	27.03	1.84		0.89	0.03	Chromite 0.64, Cu + Sn 0.08, Ni + Mn 0.78	98.46	...	or 3.0 ms 2.5 cm 0.6 / ab 10.0 di 7.6 tr 2.5 / ky 20.7 nf 39.7 / ol 21.2	Cc	J. Nuricsany	Sitzb. Wien. Akad. 1856, 20, 308-406. Mass anal. calc. by Wadsworth

ANALYSES OF STONE METEORITES—Continued

DOFEMIC, SILICOMETALLIC, PERPOLIC, PYROLIC, PERMIRIC, PERMIRIC, MAGNESIFERROUS, INCOSE

Name	SiO₂	Al₂O₃	FeO	MgO	CaO	Na₂O	K₂O	Fe	Ni	Co	S	P	Miscellaneous	Sum	Sp. gr.	Norm	Brezina's Symbol	Analyst	Reference
30. Llano del Inca	26.02	4.70	19.29	8.15	3.45	....	....	23.29	2.38	0.16	Fe S 10.61	....	Cr₂O₃ 0.29 MnO 0.06 NiO 0.90 P₂O₅ 0.70	100.00	....	an 12.8 hy 27.6 ab 1.7 ol 22.1 tr 10.6 cm 0.5 ni 25.8	M	L. G. Eakins	Proc. Rochester Acad. Sci. 1890, 1, 94. Mass anal. calc. by Farrington

DOFEMIC, SILICOMETALLIC, PERPOLIC, DOMOLIC, PERMIRIC, PERMIRIC, DOMAGNESIC, KERNOUVOSE

Name	SiO₂	Al₂O₃	FeO	MgO	CaO	Na₂O	K₂O	Fe	Ni	Co	S	P	Miscellaneous	Sum	Sp. gr.	Norm	Brezina's Symbol	Analyst	Reference
31. Kernouvé	32.95	3.19	11.70	23.68	1.89	1.41	....	22.25	1.55	....	2.15	....		100.77	3.75	ab 12.1 di 5.8 tr 6.1 an 2.2 hy 2.5 ni 33.8 ol 47.5	Ck	F. Pisani	Comptes Rendus, 1860, 68, 1480-1491

PERFEMIC, PERSILICIC, PERPOLIC, PERPYRIC, PERMIRLIC, PERMIRIC, PERMAGNESIC, BISHOPVILLOSE

Name	SiO₂	Al₂O₃	FeO	MgO	CaO	Na₂O	K₂O	Fe	Ni	Co	S	P	Miscellaneous	Sum	Sp. gr.	Norm	Brezina's Symbol	Analyst	Reference
32. Bishopville	59.97	....	....	39.34	....	....	0.74 tr	....	....	....	....	....	FeO 0.40 Li₂O tr	100.45	....	ac 0.9 mo 0.1 ns 1.2 hy 98.1	Chla	J. L. Smith	Am. Jour. Sci. 1864, 2, 38, 225

PERFEMIC, PERSILICIC, PERPOLIC, PERPYRIC, PERPYRIC, PERMIRIC, DOMAGNESIC, LBBENBÜHRENOSE

Name	SiO₂	Al₂O₃	FeO	MgO	CaO	Na₂O	K₂O	Fe	Ni	Co	S	P	Miscellaneous	Sum	Sp. gr.	Norm	Brezina's Symbol	Analyst	Reference
33. Ibbenbühren	54.49	1.06	17.34	26.12	1.22	....	....	....	....	....	....	....	MnO 0.28	100.51	3.41	an 3.1 di 2.3 hy 91.7 ol 3.5	Chl	G. von Rath	Sitzber. nieder. Gesell. Bonn, 1871, 28, 142-145
34. Manegaum	53.63	....	20.48	23.32	1.40	....	....	....	....	....	....	....	Chromite 1.03	99.95	3.20	di 6.1 cm 1.0 hy 92.0	Chl		N. S. Maskelyne, Phil. Trans. 1870, 109, 211-213
35. Shalka	52.64	....	19.78	26.38	0.55	0.40	....	....	....	....	....	....	Cr₂O₃ 0.23	99.98	3.41	ns 0.7 cm 2.2 di 2.3 hy 83.2 ol 11.4	Chl	C. Rammelsberg	Monatsber. Berlin Akad. 1870, 316-322

PERFEMIC, PERSILICIC, PERPOLIC, PERPYRIC, PERMIRLIC, DOMRIC, PERMAGNESIC, BUSTOSE

Name	SiO₂	Al₂O₃	FeO	MgO	CaO	Na₂O	K₂O	Fe	Ni	Co	S	P	Miscellaneous	Sum	Sp. gr.	Norm	Brezina's Symbol	Analyst	Reference
36. Busti	52.87	....	0.19	28.32	12.40	0.57	0.24	....	....	....	....	....	Ca S 4.13 Li₂O 0.02 Ca S O 0.44	99.18	....	ks 0.3 ak 4.1 ns 1.1 di 47.7 hy 36.6 ol 8.7	Bu	N. S. Maskelyne	Phil. Trans. 1870, 140, 193-211

PERFEMIC, PERSILICIC, PERPOLIC, DOPYRIC, PERMIRIC, PERMIRIC, PERMAGNESIC

Name	SiO₂	Al₂O₃	FeO	MgO	CaO	Na₂O	K₂O	Fe	Ni	Co	S	P	Miscellaneous	Sum	Sp. gr.	Norm	Brezina's Symbol	Analyst	Reference
37. Busti	52.73	....	4.28	37.22	1.18	....	tr	....	....	....	....	....	NiO 0.78 Na₂S 0.76 MnO 0.01 Apatite tr H₂O 0.92 CaSO₄ 1.38 Li₂O tr CaCl₂ 0.01	99.47	....	di 4.6 hy 71.0 ol 20.7	Bu	W. Dancer	Phil. Trans. 1870, 140, 193-211

ANALYSES OF STONE METEORITES—Continued

PERFEMIC, PERSILICIC, PERPOLIC, DOPYRIC, PERMIRLIC, PERMIRIC, DOMAGNESIC, SHALKOSE

Name	Si O₂	Al₂O₃	Fe O	Mg O	Ca O	Na₂ O	K₂ O	Fe	Ni	Co	S	P	Miscellaneous	Sum	Sp. gr.	Norm	Brezina's Symbol	Analyst	Reference
38. Shalka	52.51	0.66	16.81	28.35	0.89	0.22	...				Fe S 0.39	tr	Cr₂O₄ 1.25	101.08	...	ab 1.6, an 1.1, di 2.7, hy 78.4, cm 1.8, tr 0.4	Chl	H. B. von Foullon	Ann. Wien. Mus. 1888, 3, 195-208
39. Coon Butte	42.62	1.69	12.98	26.55	0.96	0.40	0.12	7.71	0.93	0.01	Fe S 2.15	Fe₃P 0.76	Fe₃O₄ 2.60, Chromite 0.08, Cu, Mn, Sn, tr	100.00	3.47	or 0.6, ab 3.8, an 2.8, di 1.5, hy 47.5, ol 20.3, ml 3.7, tr 2.8, sc 7.7	Cib	J. W. Mallet	Am. Jour. Sci. 1906, 4, 21, 353. Mass anal. calc. by Farrington

PERFEMIC, PERSILICIC, PERPOLIC, DOPYRIC, PERMIRLIC, PERMIRIC, MAGNESIFERROUS, MIDDLESBOROSE

Name	Si O₂	Al₂O₃	Fe O	Mg O	Ca O	Na₂ O	K₂ O	Fe	Ni	Co	S	P	Miscellaneous	Sum	Sp. gr.	Norm	Brezina's Symbol	Analyst	Reference
40. Ngawi	42.77	0.78	24.06	15.31	2.63	2.73	0.45	2.87	0.65	tr	Fe S 5.71	...	Chromite 0.47, Ni O 1.57, Mn O tr	100.00	...	or 2.2, ab 2.1, ns 4.9, di 10.6, cm 0.5, tr 5.7, hy 43.2, nf 3.5, ol 27.2	Ccn	E. H. von Baumhauer	Arch. Neerland, 1884, 19, 175-185
41. Middlesborough	42.61	1.75	23.80	20.86	1.60	...	...	7.22	2.00	0.16	...	...		100.00	...	an 4.7, di 2.8, nf 9.4, hy 52.2, ol 31.0	Cw	W. Flight	Phil. Trans. Roy. Soc. 1882, 3, 885-899. Mass anal. calc. by Farrington

PERFEMIC, PERSILICIC, PERPOLIC, PYROLIC, PERMIRIC, DOMAGNESIC, TRAVISOSE

Name	Si O₂	Al₂O₃	Fe O	Mg O	Ca O	Na₂ O	K₂ O	Fe	Ni	Co	S	P	Miscellaneous	Sum	Sp. gr.	Norm	Brezina's Symbol	Analyst	Reference
42. Hendersonville	46.06	2.20	14.33	28.62	2.13	0.96	0.10	2.37	0.21	0.01	1.61	0.01	Cr₂O₄ 0.23, Residue 0.51	99.35	...	or 0.6, ab 8.4, an 1.4, di 7.3, hy 30.5, ol 30.5, cm 0.2, tr 4.4, nf 2.6	Cc	Wirt Tassin	Proc. U. S. Nat. Mus. 1907, 32, 79-82
43. Travis County	44.75	2.72	16.04	27.93	2.23	1.13	0.13	1.83	0.22	0.01	1.83	...	Cr₂O₄ 0.52, Mn O tr, Cu O tr, H₂O 0.84, P₂O₅ 0.41	101.11	3.54	or 0.6, ab 9.4, an 2.0, di 5.1, hy 30.0, ol 44.0, cm 0.7, ap 5.0, nf 2.1	Cs	L. G. Eakins	Bull. U. S. Geol. Survey 1891, 78, 91
44. Ergheo	42.53	2.23	17.13	26.13	1.08	0.13	...	0.57	0.17	...	Fe S 9.48	...		99.45	3.31	ab 1.1, an 5.6, di 45.0, hy 36.8, cm 0.5, nf 0.7	Ckb	G. Boeris	Soc. d'Esploraz. Comm. in Africa, Milan 1898, 13
45. Mauerkirchen	41.53	1.71	23.32	24.20	2.12	0.24	0.15	3.75	...	...	0.70	...		98.44	...	ab 2.1, an 3.3, di 5.0, hy 32.3, ol 47.4, cm 0.7, tr 1.9, ml 3.8	Cw	F. Crook	Chem. Const. Met. Stones, 26-30
46. New Concord	40.39	2.30	18.13	23.51	2.52	...	...	5.78	0.24	...	1.46	...	Fe₂O₅ 0.82, Ni O 0.81, Mn tr	99.50	...	ab 6.4, an 4.0, ml 8.4, hy 40.6, nf 6.0, ol 33.2	Cia	A. Madelung	Buchner's Meteoriten 1863, 105
47. Sokobanja	40.14	...	25.54	25.78	...	0.26	0.06	...	...	...	1.46	tr		100.21	...	ks 0.2, ab 0.5, an 2.5, tr 4.1, nf 6.8, ol 46.0	Cc	S. M. Losanitch	Ber. Chem. Gesell. Berlin 1878, 11, 96-98. Mass anal. calc. by Wadsworth
48. Manbhoom	40.12	1.80	20.53	27.30	1.93	0.44	0.20	4.24	0.91	...	1.70	0.20	Fe₂O₅ 0.83, Cr₂O₄ 0.55, Mn O 0.07	100.82	...	or 1.1, ab 3.7, an 2.5, di 5.6, hy 24.8, ol 48.9, ml 1.2, cm 0.0, tr 4.7, sc 1.2	Ru	H. B. von Foullon	Ann. Wien. Mus. 1888, 3, 195-208
49. Long Island	35.65	3.08	22.85	22.74	1.40	0.25	0.03	2.60	0.67	0.04	1.90	0.06	Cr₂O₄ 6.33, Ni O 0.77, Co O 0.06, Mn O tr, H₂O 1.52	99.95	3.45	ab 2.1, an 7.0, C 0.1, hy 27.3, ol 42.3, cm 0.4, tr 5.2, sc 0.4, nf 3.3	Cia	H. W. Nichols	Pubs. Field Col. Mus. Geol. Ser. 1902, 1, 297

PERFEMIC, PERSILICIC, PERPOLIC, PYROLIC, PERMIRIC, MAGNESIFERROUS, CONCORDOSE

Name	Si O₂	Al₂O₃	Fe O	Mg O	Ca O	Na₂ O	K₂ O	Fe	Ni	Co	S	P	Miscellaneous	Sum	Sp. gr.	Norm	Brezina's Symbol	Analyst	Reference
50. New Concord	41.73	0.28	24.72	21.64	0.02	0.92	...	9.23	1.31	0.04	0.11	tr	Cu tr, Mn tr	100.00	3.55	ab 1.6, an 1.4, ns 1.5, hy 50.4, ol 35.6, nf 10.6	Cia	J. L. Smith	Am. Jour. Sci. 1861, 2, 31, 87-98. Mass anal. calc. by Farrington

ANALYSES OF STONE METEORITES—Continued

PERFEMIC, PERSILICIC, PERPOLIC, DOMOLIC, PERMIRLIC, PERMIRIC, DOMAGNESIC, WACONDOSE

Name	SiO_2	Al_2O_3	FeO	MgO	CaO	Na_2O	K_2O	Fe	Ni	Co	S	P	Miscellaneous	Sum	Sp. gr.	Norm	Brezina's Symbol	Analyst	Reference
51. Zavid	41.90	1.92	27.40	22.79	4.60	1.05	0.41	0.15	.....	.....	1.01	.....	H_2O 0.30	101.11	3.55	or 2.2 / ab 7.9 / ns ... / di 18.7 / tr 2.7 / nf 0.2	Cia	C. Hödlmoser	Wiss. Mitth. Bosnia u. Herzegovinia, 1901, 8, 410
52. Nowo-Urei	39.51	0.60	13.35	35.80	1.40	.....	.....	5.25	0.20	.....	0.15	0.02	Cr_2O_3 0.05, MnO 0.43, Carbon 1.26, Diamond 1.00	99.92	...	am 1.7 / di 4.2 / hy 16.7 / tr 0.4 / ol 67.2 / nf 5.5	U	M. Jerofejeff and P. Latschinoff	Verh. d. Russ. Kais. Minor. Ges. 1888, 24, 34 pp.
53. Cynthiana	38.99	0.22	19.73	26.56	2.20	0.49	.....	5.36	0.50	0.07	Fe S 5.50	.....	Cr_2O_3 0.15	99.77	3.41	ab 1.1 / ns 0.7 / di 8.8 / hy 23.6 / tr 5.5 / ol 54.6 / nf 3.9	Cg	J. L. Smith	Am. Jour. Sci. 1877, 3, 14, 226. Mass anal. calc. by Wadsworth
54. Waconda	38.14	1.02	23.44	26.69	tr	1.05	tr	4.64	0.65	0.05	Fe S 3.85	tr	MnO 0.47, Li_2O tr, Cu tr	100.00	3.50	ab 5.2 / ns 1.1 / hy 14.9 / tr 3.9 / ol 69.0 / nf 5.3	Ccb	J. L. Smith	Am. Jour. Sci. 1877, 3, 13, 212. Mass anal. calc. by Farrington
55. Bluff	37.70	2.17	23.82	25.94	2.20	.....	.....	4.41	0.88	0.37	1.30	.....	NiO 1.50, CoO 0.16, MnO 0.45, P_2O_5 0.25	101.24	3.51	am 6.1 / di 2.5 / hy 19.4 / ap 0.7 / tr 3.6 / ol 63.0 / nf 5.7	Ck	J. E. Whitfield	Am. Jour. Sci. 1888, 3, 36, 119

PERFEMIC, PERSILICIC, PERPOLIC, DOMOLIC, PERMIRLIC, PERMIRIC, MAGNESIFERROUS, KABOSE

Name	SiO_2	Al_2O_3	FeO	MgO	CaO	Na_2O	K_2O	Fe	Ni	Co	S	P	Miscellaneous	Sum	Sp. gr.	Norm	Brezina's Symbol	Analyst	Reference
56. Chateau Renard	38.13	3.82	29.44	17.67	0.14	0.86	0.27	7.70	1.55	.....	0.39	.....	Chromite 0.80, MnO 0.05, Co 0.58	99.97	3.56	or 1.7 / ns 7.3 / hy 24.9 / tr 1.1 / ol 52.9 / nf 9.3 / C 1.7 / C 0.8	Cia	A. Dufrenoy	Comptes Rendus, 1841, 13, 47 53
57. Kaba	34.24	5.38	26.20	22.39	0.66	.....	0.30	2.88	1.37	tr	Fe S 3.55	tr	MnO 0.05, Co 0.01	98.50	...	am 3.3 / ky 15.0 / tr 3.6 / ol 65.4 / nf 4.3 / C 3.9	K	F. Wohler	Sitzber. Wien. Akad. 1858, 33, 205 209

PERFEMIC, PERSILICIC, PERPOLIC, PEROLIC, PERMIRIC, PERMIRIC, DOMAGNESIC, KAKOVOSE

Name	SiO_2	Al_2O_3	FeO	MgO	CaO	Na_2O	K_2O	Fe	Ni	Co	S	P	Miscellaneous	Sum	Sp. gr.	Norm	Brezina's Symbol	Analyst	Reference
58. Kakova	37.97	2.27	22.68	24.98	0.69	1.77	0.52	7.15	1.24	0.09	.....	0.01	Chromite 0.07, MnO 0.42, Graphite 0.14	100.00	3.38	or 2.8 / ab 9.4 / ns 1.2 / di 0.4 / tr ... / ol 76.0 / nf 8.5 / am 0.9	Cga	E. P. Harris	Chem. Const. 1859, 22-34. Met. Mass anal. calc. by Farrington

PERFEMIC, PERSILICIC, PERPOLIC, PEROLIC, PERMIRLIC, PERMIRIC, MAGNESIFERROUS, JEROMOSE

Name	SiO_2	Al_2O_3	FeO	MgO	CaO	Na_2O	K_2O	Fe	Ni	Co	S	P	Miscellaneous	Sum	Sp. gr.	Norm	Brezina's Symbol	Analyst	Reference
59. Warrenton	35.51	0.13	30.17	25.57	1.43	0.23	.....	1.79	0.21	.....	Fe S 3.51	.....	Cr_2O_3 0.06, NiO 0.16, CoO 0.23	100.00	3.47	ab 0.5 / ns 0.2 / di 5.7 / ky 3.5 / tr 3.5 / ol 85.4 / nf 2.0	Cco	J. L. Smith	Am. Jour. Sci. 1877, 3, 14, 223. Mass anal. calc. by Farrington
60. Felix	33.57	3.24	26.22	19.74	5.45	0.62	0.14	2.59	0.36	0.08	Fe S 4.76	.....	Cr_2O_3 0.80, NiO 0.01, MnO 0.68, H_2O 0.16, CaO 0.01, Graphite 0.36	99.79	3.78	k 0.4 / am 2.8 / di 0.8 / kp 4.8 / am 5.6 / tr 7.7 / nf 3.0	Kc	Peter Fireman	Proc. U. S. Nat. Mus. 1901, 24, 193-198
61. Jerome	33.11	1.77	27.97	21.59	1.31	0.65	0.28	3.81	0.43	0.01	1.88	.....	Cr_2O_3 0.58, NiO 1.77, H_2O 3.53, P_2O_5 0.37	98.56	3.47	or 1.7 / ab 5.8 / di 2.3 / ns 0.9 / tr 1.0 / ol 72.7 / am 1.1 / nf 4.3	Cck	H. S. Washington	Am. Jour. Sci. 1898, 4, 5, 453

223

ANALYSES OF STONE METEORITES—*Continued*

PERFEMIC, PERSILICIC, DOPOLIC, PYROLIC, PERMIRIC, PERMIRIC, DOMAGNESIC, ELWAHOSE

Name	SiO₂	Al₂O₃	Fe O	Mg O	Ca O	Na₂O	K₂O	Fe	Ni	Co	S	P	Miscellaneous	Sum	Sp.gr.	Norm	Brezina's Symbol	Analyst	Reference
62. Eli Elwah	39.47	2.87	17.06	25.58	1.61	0.73	0.11	....	1.01	....	2.30	0.10	Fe₂O₃ 9.18	100.02	....	or 0.6 di 2.0 ml 13.5 / ab 5.8 hy 34.6 tr 0.3 / an 4.4 ol 20.1 sc 0.6 / nf 1.0	C	A. Liversidge	Proc. Roy. Soc. New South Wales, 1903, 341–359

PERFEMIC, DOSILICIC, PERPOLIC, PERPYRIC, PERMIRIC, PERMAGNESIC, HVITTISOSE

Name	SiO₂	Al₂O₃	Fe O	Mg O	Ca O	Na₂O	K₂O	Fe	Ni	Co	S	P	Miscellaneous	Sum	Sp.gr.	Norm	Brezina's Symbol	Analyst	Reference
63. Bremervörde	45.40	2.34	4.36	22.40	....	1.18	0.37	21.61	1.89	....	....	....	Chromite 0.31 Graphite 0.14	100.00	3.54	or 2.2 hy 63.7 cm 0.3 / ab 10.0 ol 0.3 nf 23.5	Ccb	F. Wöhler	Ann. Chem. Pharm. 1856, 99, 244–248
64. Hvittis	41.53	1.55	0.34	23.23	1.41	1.26	0.32	24.66	1.96	0.07	3.30	0.08	Cr₂O₃ 0.34	100.28	....	or 1.7 ms 0.0 cm 0.9 / ab 6.8 di 5.4 tr 0.6 / ol 52.2 sc 0.6 / ol 2.4 nf 20.4	Cek	L. H. Borgström	Die Meteoriten von Hvittis u. Marjalahti, Helsingfors 1903, 24

PERFEMIC, DOSILICIC, PERPOLIC, PERPYRIC, PERMIRLIC, DOMAGNESIC, MOCSOSE

Name	SiO₂	Al₂O₃	Fe O	Mg O	Ca O	Na₂O	K₂O	Fe	Ni	Co	S	P	Miscellaneous	Sum	Sp.gr.	Norm	Brezina's Symbol	Analyst	Reference
65. Mocs	42.74	tr	20.86	15.95	2.78	1.20	0.21	7.93	1.38	tr	2.61	0.41	Chromite 1.56 MnO 0.57 / MnO 1.12 Li₂O tr / C 0.19	99.51	3.64	Q 3.5 kt 0.3 cm 1.6 / or 2.3 tr 7.1 / hy 58.8 nf 9.9	Cwa	F. Koch	Min. Mitth. 1883, 2, 5, 243
66. St. Mark's	38.29	0.64	6.50	18.23	1.08	0.85	0.23	26.44	1.84	0.21	5.26	0.05	MnO 0.33 Cl 0.27 / Mn 0.20 C 0.36 / Ca 0.28	101.15	....	Q 1.2 tr 14.2 / or 1.1 di 4.5 ok 0.4 / ab 2.1 hy 56.0 nf 19.4	Ck	E. Cohen	Ann. South African Mus. 1906, 5, 1–16

PERFEMIC, DOSILICIC, PERPOLIC, DOPYRIC, PERMIRLIC, PERMIRIC, DOMAGNESIC, CASTALIOSE

Name	SiO₂	Al₂O₃	Fe O	Mg O	Ca O	Na₂O	K₂O	Fe	Ni	Co	S	P	Miscellaneous	Sum	Sp.gr.	Norm	Brezina's Symbol	Analyst	Reference
67. Modoc	44.13	2.47	15.37	26.45	1.74	0.44	tr	6.56	0.68	0.03	1.38	0.05	MnO 0.10	99.40	3.54	ab 3.7 di 2.0 cm 3.8 / an 5.0 hy 47.4 sc 0.2 / ol 28.4 nf 7.3	Cwa	Wirt Tassin	Am. Jour. Sci. 1906, 4, 21, 359
68. Krähenberg	41.78	0.06	19.53	24.44	1.94	1.00	....	6.31	0.54	....	2.17	....	Chromite o. MnO tr	98.68	3.50	ab 0.5 di 1.8 tr 6.1 / hy 7.6 nf 6.9 / ol 26.8	Cho	G. von Rath	Ann. Phys. 1866, 137, 328–336. Mass anal. calc. by Wadsworth
69. Bachmut	39.59	2.71	18.81	23.37	0.04	0.63	tr	8.52	1.24	....	2.37	0.05	Chromite 0.79 MnO 0.04 MnO 0.21	98.37	3.56	ab 5.2 hy 26.2 cm 0.8 / C 1.6 ol 26.6 6.5 / nf 10.0	Cw	A. Kühlberg	Archiv. Nat. Liv. Ehst. Kurlands 1867, 1, 4, 132
70. Drake Creek	38.50	4.81	10.03	22.79	0.70	0.59	0.02	12.82	1.50	0.16	1.80	....	Cr₂O₃ 1.37 NiO, CuO, SnO₂ 2.53 Cu + Sn 0.07	100.00	....	ab 5.2 hy 43.1 cm 2.0 / an 3.6 ol 23.6 tr 4.9 / C 2.5 nf 14.6	Cwa	E. H. Baumhauer	Ann. Phys. Chem. 1845, 66, 498–503
71. Castalia	38.50	2.14	13.31	29.83	....	0.55	tr	14.19	0.96	0.06	0.46 Fe S	....	Li₂O tr	100.00	....	ab 4.7 hy 27.5 tr 1.2 / an 1.2 ol 40.9 nf 15.3	Cgb	J. L. Smith	Am. Jour. Sci. 1875, 3, 10, 147–148
72. Dundrum	37.80	0.85	7.92	23.33	1.32	0.96	0.50	19.57	1.03	....	4.05	....	Chromite 1.50 MnO 0.16	98.99	3.32	or 0.6 di 1.5 cm 1.5 / ab 4.2 hy 5.1 tr 4.1 / ol 20.8 nf 20.6 / ol 21.4	Ck	S. Haughton	Proc. Roy. Soc. 1866, 15, 214–217. Mass anal. calc. by Wadsworth
73. Gopalpur	37.44	2.52	11.94	19.72	1.60	0.62	0.21	20.96	1.80	0.10	1.74	....	Cr₂O₃ tr MnO 0.26	98.91	....	or 1.1 di 3.6 tr 4.8 / ab 5.2 hy 41.9 nf 22.9 / an 3.6 ol 15.1	Cc	A. Exner	Min. Mitth. 1872, 41–43

ANALYSES OF STONE METEORITES—Continued

PERFEMIC, DOSILICIC, PERPOLIC, DOPYRIC, PERMIRIC, PERMIRLIC, DOMAGNESIC, CASTALIOSE—Continued

Name	SiO_2	Al_2O_3	FeO	MgO	CaO	Na_2O	K_2O	Fe	Ni	Co	S	P	Miscellaneous	Sum	Sp. gr.	Norm	Brezina's Symbol	Analyst	Reference
74. Adare	37.26	2.03	8.95	13.50	3.61	0.79	0.12	16.24	2.73	0.10	6.54 (Fe S)	....	Chromite 1.75, MnO 5.50, V tr	99.12	3.93	or 0.6 di 13.2 cm 1.8, ab 6.8 hy 37.7 tr 6.5, an 1.7 ol 11.8 nf 19.1	Cga	R. Apjohn	Jour. Chem. Soc. 1874, 2, 12, 104–106. Mass anal., calc. by Wadsworth
75. Tokuchimura	36.34	....	14.76	20.91	2.47	1.18	0.28	16.58	1.82	0.05	2.75	0.08	Fe O 0.36 Mn O 0.16, Cr O 0.42 NiO 0.30, Chromite 0.95	99.40	3.81	kt 0.5 ml 0.7, ns 2.3 cm 1.6, di 9.8 tr 7.6, hy 42.5 sc 0.6, ol 14.0 nf 18.5	Ck	Lindner	Ber. Berlin Akad. 1904, 978–983
76. Stalldalen	35.71	2.11	10.29	23.16	1.61	0.62	0.15	21.10	1.61	0.17	2.27	0.01	Cr₂O₃ 0.40 P₂O₅ 0.30, NiO 0.20 Cl 0.04, MnO 0.25	100.00	3.74	or 0.6 di 41.3 ap 0.7, ab 5.2 ol 20.3 tr 6.3, C 1.1 nf 22.9	Cga	G. Lindström	Öfversigt. Kongl. Veten. Forhan. 1877, 35
77. Gnadenfrei	32.11	1.60	14.88	17.03	2.01	0.70	....	25.16	3.92	tr	1.87	....	Cr₂O₃ 0.57, Mn O tr, P₂O₅ tr	99.85	3.71	ab 5.8 di 7.0 cm 0.9, 1.4 ol 22.4 tr 5.1, nf 29.1	Cc	Galle and Lasaulx	Monatsber. Berlin Akad. 1879, 750–771
78. Orgueil	26.08	0.90	15.77	17.00	1.85	2.26	0.19	....	....	....	13.43 (Fe S)	....	Fe₂O₃ 7.78, Chromite 0.40, MnO 0.36, H₂O and org. matter 13.80	100.00	2.50	or 1.1 ml 11.4, ab 3.7 di 7.5 cm 0.5, ol 44.6 tr 13.4	K	Pisani	Comptes Rendus 1864, 59, 134

PERFEMIC, DOSILICIC, PERPOLIC, DOPYRIC, PERMIRLIC, PERMIRIC, MAGNESIFERROUS, ENSISHEIMOSE

Name	SiO_2	Al_2O_3	FeO	MgO	CaO	Na_2O	K_2O	Fe	Ni	Co	S	P	Miscellaneous	Sum	Sp. gr.	Norm	Brezina's Symbol	Analyst	Reference
79. Ensisheim	35.65	2.31	34.19	13.13	1.78	0.38	0.22	8.00	1.23	....	2.05	1.01	Cr₂O₃ 0.41, Mn O 0.21	99.57	3.50	or 1.1 di 3.0 cm 0.7, ab 3.1 hy 38.8 tr 5.6, an 4.2 ol 25.2 sc 6.2, nf 9.2	Ckb	F. Crook	Chem. Const. Met. Stones, 21–26

PERFEMIC, DOSILICIC, PERPOLIC, PYROLIC, PERMIRLIC, PERMIRIC, PERMAGNESIC, ORVINIOSE

Name	SiO_2	Al_2O_3	FeO	MgO	CaO	Na_2O	K_2O	Fe	Ni	Co	S	P	Miscellaneous	Sum	Sp. gr.	Norm	Brezina's Symbol	Analyst	Reference
80. Orvinio	37.42	2.27	7.98	22.90	2.32	1.21	0.29	22.23	2.60	....	1.99	....	Cr₂O₃ 0.62, MnO 0.07, Sn 0.08	101.19	3.64	or 1.7 di 8.7 tr 5.5, ab 10.0 hy 24.6 nf 24.8, an 0.3 ol 20.4	Co	L. Sipöcz	Sitzber. Wien Akad. 1875, 52, 1, 464
81. Klein-Wenden	33.03	3.75	6.90	23.64	2.83	0.28	0.38	23.90	2.37	....	2.09	0.02	Cr₂O₃ 0.62, MnO 0.07	100.01	3.70	or 2.2 di 4.8 cm 0.9, ab 2.1 ol 20.9 tr 5.8, an 8.1 27.8 nf 26.4	Ck	C. Rammelsberg	Ann. Phys. Chem. 1844, 62, 449–464

PERFEMIC, DOSILICIC, PERPOLIC, PYROLIC, PERMIRLIC, PERMIRIC, DOMAGNESIC, PULTUSKOSE

Name	SiO_2	Al_2O_3	FeO	MgO	CaO	Na_2O	K_2O	Fe	Ni	Co	S	P	Miscellaneous	Sum	Sp. gr.	Norm	Brezina's Symbol	Analyst	Reference
82. Pultusk	41.54	1.17	14.04	26.73	0.28	1.34	....	11.51	0.65	....	0.87	tr	Chromite 0.29, Mn O 0.40, Insol. 0.04	99.01	3.66	ab 6.3 ns 1.2 cm 0.3, hy 38.6 tr 2.4, ol 40.5 nf 12.2	Cga	G. von Rath	Neues Jahrb. Min. 1869, 80–82. Mass anal., calc. by Wadsworth
83. Searsmont	40.82	0.81	13.84	25.99	....	0.85	....	13.24	1.33	0.06	3.06 (Fe S)	....	Chromite tr, Li₂O tr	100.00	3.70	ab .2 ns tr 3.1, di 44.0 nf 14.6, ol 33.4	Cc	J. L. Smith	Am. Jour. Sci. 1871, 3, 2, 200. Mass anal. calc. by Farrington
84. Rochester	40.77	0.10	16.52	26.47	2.43	0.58	....	9.52	0.42	0.05	2.99 (Fe S)	....	Chromite 0.15	100.00	3.55	ab 0.5 di 1.0 cm 0.2, hy 33.6 tr 3.0, ol 42.1 nf 10.0	Cc	J. L. Smith	Am. Jour. Sci. 1877, 3, 14, 222. Mass anal. calc. by Farrington

ANALYSES OF STONE METEORITES—Continued

PERFEMIC, DOSILICIC, PERPOLIC, PYROLIC, PERPOLIC, PERMIRLIC, PERMIRIC, DOMAGNESIC, PULTUSKOSE—Continued

Name	Si O₂	Al₂O₃	Fe O	Mg O	Ca O	Na₂O	K₂O	Fe	Ni	Co	S	P	Miscellaneous	Sum	Sp. gr.	Norm	Brezina's Symbol	Analyst	Reference
85. Dhurmsala	40.69	0.60	11.20	26.59	…	0.39	0.21	6.88	1.54	…	Fe S 5.61	…	Chromite 4.16 / Mn O 1.26	99.13	3.40	or 1.1 ns 0.2 cm 4.2 / ab 2.1 ky 16.9 tr 5.6 / di 39.4 nf 8.4	Ci	S. Haughton	Proc. Roy. Soc. 1866, 15, 214-217. Mass anal. calc. by Wadsworth
86. Richmond	40.37	2.21	13.82	28.33	2.68	…	…	8.22	…	…	Fe S 4.37	…		100.00	3.37	an 6.1 di 5.6 tr 4.4 / hy 30.2 nf 8.2 / ol 45.5	Cck	C. Rammelsberg	Monatsber. Berlin Akad. 1870, 70, 440
87. Tieschitz	40.23	1.93	19.80	20.55	1.54	1.53	…	10.26	1.31	…	1.65	…		98.80	3.59	ab 0.9 ns 0.6 tr 4.5 / di 6.3 nf 11.6 / hy 39.0 / ol 34.1	Cc	J. Habermann	Denkschr. Wien Akad. 1879, 39, 187-201
88. St. Denis-Westrem	40.20	2.54	16.22	25.08	2.00	0.99	tr	10.37	1.24	0.12	2.12	…	Cr₂O₃ 0.90 / Mn O tr	101.78	…	ab 8.4 di 6.0 cm 1.4 / an 2.5 ky 26.7 tr 5.8 / ol 38.3 nf 11.7	Cca	C. Klement	Bull. Mus. roy. d'hist. Nat. Belgique 1886, 4, 280
89. St. Christophe	39.33	2.15	13.66	25.90	1.51	0.51	0.18	7.79	1.67	0.11	Fe S 6.90	…	Cr₂O₃ 0.38	100.09	…	or 1.1 di 3.8 cm 0.0 / ab 4.2 ky 27.0 tr 0.6 / an 3.1 hy 42.8 nf 9.6	Cg	M. A. Lacroix	Bull. Soc. de'l Onest de la France, 1906, 2, 6, 81-112
90. Tadjera	39.20	1.64	14.18	25.68	2.66	…	…	8.32	…	…	Fe S 8.04	…	Cr₂O₃ 0.12	99.84	3.60	an 4.5 di 6.0 cm 0.2 / ky 33.4 tr 8.0 / ol 38.4 nf 8.3	Ct	S. Meunier	Comptes Rendus 1868, 66, 513-519
91. Shelburne	39.19	2.15	15.16	26.24	1.75	0.73	0.22	10.70	0.78	0.04	1.61	0.06	Cr₂O₃ 0.62 / Mn O 0.12	99.37	3.50	or 1.1 di 4.9 cm 0.0 / ab 5.8 ky 25.5 tr 4.4 / an 2.5 ol 41.6 nf 11.5	Cg	L. H. Borgström	Trans. Roy. Astr. Soc. of Canada 1904
92. Alfianello	39.14	0.93	17.42	25.01	1.96	0.75	0.10	11.31	1.09	…	2.71	…		100.42	…	or 0.6 di 0.5 tr 7.4 / ab 4.2 ky 3.8 nf 12.4 / hy 37.7 / ol 37.5	Ci	H. von Foullon	Sitzber. Wien Akad. 1883, 88, 1, 433. Monatsber. Berlin Akad. 1870, 457-459. Mass anal. calc. by Wadsworth
93. Marion	38.96	2.00	14.52	26.05	1.18	0.38	tr	13.51	1.08	…	Fe S 2.32	Fe₃P 2.00		100.00	…	ab 3.7 di 1.5 tr 6.3 / hy 41.8 nf 14.6 / ol 28.4	Cwa	C. Rammelsberg	Comptes Rendus 1859, 49, 31-36
94. Aussun	38.72	1.85	16.93	22.53	0.80	0.57	0.11	8.63	0.96	…	Fe S 3.74	…	Chromite 1.83 / Mn O tr	98.67	3.54	or 0.6 di 1.4 cm 1.8 / ab 4.7 hy 35.2 tr 3.7 / an 2.2 ol 37.3 sc 2.0 nf 9.6	Cc	H. A. Damour	Am. Jour. Sci. 1894, 3, 47, 430. Mass anal. calc. by Farrington
95. Beaver Creek	37.43	2.17	10.49	23.73	1.76	0.80	0.09	15.53	1.51	0.08	Fe S 5.05	…	Magnetite 0.16 H₂O 0.20 / Chromite 0.30 TiO₂ 0.08 / NiO 0.03 Cu 0.01 / MnO 0.24 P₂O₅ 0.25	100.00	…	or 0.6 di 4.5 cm 0.3 / ab 6.8 ky 35.0 tr 0.3 / an 2.2 ol 36.2 ap 0.3 sc 5.1 nf 17.1	Cck	W. F. Hillebrand	Private contribution
96. Saline	37.08	1.83	18.04	23.34	2.03	0.26	0.08	7.89	0.95	0.04	1.65	0.05	Fe₂O₃ 4.45 H₂O 1.23 / Cr₂O₃ 1.25 / NiO 0.74 / CoO 0.07	100.99	3.62	or 0.6 di 5.3 mt 6.5 / ab 2.1 ky 32.0 cm 2.0 / an 3.6 ol 33.2 sc 0.2 tr 4.5 nf 8.9	Cck	H. W. Nichols and E. W. Tillotson	Private contribution
97. Hessle	36.83	2.38	10.85	23.21	1.80	0.94	…	20.08	2.15	0.02	1.88	0.15	Cr₂O₃ 0.07 / Mn O 0.42 / Cl 0.04	100.84	3.70	ab 7.9 di 5.1 tr 5.1 / an 2.5 ky 28.8 sc 0.8 / hy 27.3 nf 22.3	Cc	G. Lindström	Kongl. Svenske. Vet. Ak. 1870
98. Ogi	36.75	1.89	8.84	23.36	1.94	0.97	0.16	15.35	…	1.75	Fe S 5.91	…	Chromite 0.61 Cu+ / NiO 0.30 Sn 0.15 / MnO 0.51 Mn 0.18 / P₂O₅ 0.34	99.01	…	or 1.1 di 6.0 cm 0.6 / ab 8.4 ky 32.7 ap 0.7 / an 0.3 ol 35.0 nf 17.4	Cw	T. Shimidzu	Trans. Asiatic Soc. Japan 1882, 10, 199-203
99. Lixna	36.45	2.52	13.16	25.08	tr	0.72	tr	16.95	1.71	…	2.13	0.14	Chromite 0.70 / Mn O 0.03 / … 0.43	100.02	3.73	ab 5.8 di 30.0 tr 0.7 / C 1.4 ky 26.7 tr 5.9 / nf 19.1	Cga	A. Kuhlberg	Archiv. Nat. Liv. Ehst. Kurlands 1867, 1, 4, 1-32

226

PERFEMIC, DOSILICIC, PERPOLIC, PYROLIC, PERMIRLIC, PERMIRIC, DOMAGNESIC, PULTUSKOSE — *Continued*

Name	SiO₂	Al₂O₃	FeO	MgO	CaO	Na₂O	K₂O	Fe	Ni	Co	S	P	Miscellaneous	Sum	Sp. gr.	Norm	Brezina's Symbol	Analyst	Reference
100. Salt Lake City	36.05	1.96	11.70	23.02	1.87	0.85	0.06	15.67	1.38	0.10	Fe S 5.51		Chromite 0.62, H₂O 0.04, P₂O₅ 0.26	100.00	3.66	or 0.6, ab 7.3, an 1.4; di 4.9, hy 22.8, ol 37.7; cm 0.6, ab 0.7, tr 5.5, nf 17.2		S. L. Penfield	Am. Jour. Sci. 1886, 3, 32, 228
101. Pultusk	35.85	1.96	12.12	24.95	1.56	0.95	0.39	15.55	2.21				Fe₂O₃ 3.85	99.39		or 2.2, ab 3.1; ns 1.1, hy 6.2, ol 21.7; m 5.5, nf 17.8	Cga	C. Rammelsberg	Monatsber. Berlin Akad. 1870, 448–452. Mass anal. calc. by Wadsworth
102. Khetree	35.17	1.77	11.16	23.80	2.37	0.87	tr	18.79	1.26	0.21	Fe S 1.76	0.12	Cr₂O₃ 0.40, Cr₂O₃ 0.10	97.78	3.68	ab 7.3, an 1.1; di 8.4, hy 20.4, ol 33.2; cm 0.7, sc 4.8, nf 20.4	Cgb	D. Waldie	Jour. Asiat. Soc. Bengal 1869, 38, 2, 252–258
103. Allegan	34.95	2.55	8.47	21.99	1.73	0.66	0.23	21.09	1.81	0.15	Fe S 5.05		Cr₂O₃ 0.53, H₂O 0.25, NiO tr, TiO₂ 0.02, MnO 0.18, Cu 0.01, Li₂O tr, P₂O₅ 0.27	100.00	3.91	or 1.1, ab 5.8; di 2.4, hy 27.7, ol 29.8; cm 0.7, il 0.2, tr 0.7, nf 23.1	Cco	H. N. Stokes	Proc. Washington Acad. Sci. 1900, 2, 41

PERFEMIC, DOSILICIC, PERPOLIC, PYROLIC, PERMIRLIC, PERMIRIC, MAGNESIFERROUS, HOMESTEADOSE

Name	SiO₂	Al₂O₃	FeO	MgO	CaO	Na₂O	K₂O	Fe	Ni	Co	S	P	Miscellaneous	Sum	Sp. gr.	Norm	Brezina's Symbol	Analyst	Reference
104. Homestead	36.98	1.18	22.39	18.21	1.39	0.82	0.57	10.27	2.05		Fe S 5.25		Cr₂O₃ 0.49, MnO 0.25	99.85	3.75	or 3.3, ab 3.1; ns 0.0, hy 5.7, ol 42.1; tr 5.3, nf 12.3	Cgb	Gümber and Schwager	Sitzber. München Akad. 1875, 5, 313–339. Mass anal. calc. by Wadsworth
105. Homestead	36.92	0.64	22.64	20.02		1.42		11.17	1.30	0.07	Fe S 5.82		Li₂O tr	100.00	3.57	ab 3.1; ns 2.1, hy 34.9, ol 41.5; tr 5.8, nf 12.5	Cgb	J. L. Smith	Am. Jour. Sci. 1875, 3, 10, 362. Mass anal. calc. by Farrington

PERFEMIC, DOSILICIC, PERPOLIC, DOPOLIC, PERMIRLIC, PERMIRIC, DOMAGNESIC, FARMINGTONOSE

Name	SiO₂	Al₂O₃	FeO	MgO	CaO	Na₂O	K₂O	Fe	Ni	Co	S	P	Miscellaneous	Sum	Sp. gr.	Norm	Brezina's Symbol	Analyst	Reference
106. Lumpkin	40.73	2.28	14.70	28.10	0.04	1.05		6.11	0.84	0.05	Fe S 6.10		Cr₂O₃ 0.58, NiO 0.31, Cr₂O₃ tr, MnO 0.16	100.00	3.65	ab 8.9, C 0.5; ky 26.2, al 51.1; tr 6.1, nf 7.0	Cck	J. L. Smith	Am. Jour. Sci. 1870, 2, 50, 339. Mass anal. calc. by Farrington
107. Farmington	39.95	1.79	15.77	26.16	1.75	0.73	0.11	6.68	0.94	0.06	Fe S 5.00		Cr₂O₃ 0.66, MnO + NiO 0.61, CuO + SnO 0.35, Cu + SnO 0.02	100.00		or 0.6, ab 5.8, an 1.7; di 5.6, hy 23.1, ol 40.7; cm 0.0, tr 5.0, nf 7.7	Cs	L. G. Eakins	Am. Jour. Sci. 1892, 3, 43, 66. Mass anal. calc. by Farrington
108. Utrecht	39.30	2.25	15.30	24.37	1.48	1.39	0.15	11.07	1.24		1.90	0.01	Cr₂O₃ 0.77, MnO 0.30, Cu + SnO 0.24, FeS 2.53	100.00	3.61	or 1.1, ab 11.0; ns 0.2, hy 6.0, ol 43.8; cm 0.9, tr 5.2, nf 12.3	Cca	F. H. Baumhauer	Ann. Phys. Chem. 1845, 66, 465–498
109. Aussun	38.79	2.27	18.15	25.29		1.14	0.18	7.11	1.02	0.06	2.11		Cr₂O₃ 0.77, MnO 0.30, Cu + SnO 0.24, MnO 0.4, FeS 2.53	100.00	3.50	or 1.1, ab 0.4, C 0.3; ky 25.2, ol 45.0; cm 1.1, tr 8.3, nf 3.5	Cc	E. P. Harris	Chem. Const. Meteorites 1859, 44–51. Mass anal. calc. by Farrington
110. Mauerkirchen	38.14	2.51	25.70	21.73	2.27	1.00	0.48	6.30			2.09	0.14	Cr₂O₃ 0.39	100.75	3.46	or 2.8, ab 8.4, C 1.1; di 8.2, hy 8.5, cm 0.7; sc 1.0, nf 6.3	Cw	A. Schwager	Sitzber. München Akad. 1878, 8, 16–24
111. Alfianello	37.63	1.78	24.44	23.43	0.89	1.09	0.24	5.76	1.14	0.08	2.54	0.15	Cr₂O₃ 0.10, MnO 0.13, Cr₂O₃ 0.62	100.00		or 1.1, ab 8.3; di 3.6, hy 17.3, ol 51.9; cm 1.0, tr 7.0, nf 7.0	Ci	P. Maissen	Gazetta Chemica 1884, 13, 369
112. Blansko	37.08	2.39	14.95	23.90	1.25	0.74	0.19	16.09	0.87	0.06	0.06		Chromite 0.62, N O 0.21, MnO 0.40, Cu + SnO 0.08	98.98	3.40	or 1.1, ab 6.3, an 2.8; di 2.7, hy 18.4, ol 50.5; tr 0.2, nf 17.1	Cga	J. J. Berzelius	Ann. Phys. Chem. 1834, 33, 8–25. Mass anal. calc. by von Reichenbach 1865, 124, 213

ANALYSES OF STONE METEORITES—Continued

PERFEMIC, DOSILICIC, PERPOLIC, DOMOLIC, PERMIRLIC, PERMIRIC, DOMAGNESIC, FARMINGTONOSE—Continued

Name	Si O₂	Al₂ O₃	Fe O	Mg O	Ca O	Na₂ O	K₂ O	Fe	Ni	Co	S	P	Miscellaneous	Sum	Sp. gr.	Norm	Brezina's Symbol	Analyst	Reference
113. Hessle	36.91	1.55	13.43	25.06	2.08	1.57	....	16.36	2.16	tr	0.18	tr	Cu + Sn 0.02 Cu O 0.68	100.00	3.92	ab 7.9 ns 1.2 tr 0.5 di 8.5 nf 18.5 ol 50.0	Cc	A. E. Nordenskjöld	Ann. Phys. Chem 1870, 141, 205–224
114. Buschhof	36.01	2.48	20.98	27.17	0.71	0.26	0.33	7.92	1.51	tr	2.18	0.01	C + Sn O₂ + loss 0.15	100.00	3.52	or 1.7 ky 17.5 cm 0.2 an 3.6 ol 57.7 tr 6.0 nf 9.4	Cwa	Grewingk and Schmidt	Archiv. Nat. Liv. u. Ehst. Kurlands 1864, 3, 421–554
115. Forest City	35.62	2.08	10.27	23.93	1.40	0.81	0.06	18.08	1.19	0.13	Fe S 6.19	tr	Cr₂O₃ 0.10 P₂O₅ tr NiO 0.14 MnO tr	100.00	3.64	or 0.6 di 4.0 cm 0.2 ab 0.8 ol 40.2 tr 0.7 an 2.0 nf 19.4	Ccb	L. G. Eakins	Am. Jour. Sci. 1890, 3, 40, 320. Mass anal. calc. by Farrington
116. Cape Girardeau	35.57	2.27	11.04	23.75	1.38	0.86	0.11	16.46	1.32	0.11	Fe S 5.68	tr	Chromite 0.68 H₂O 0.47 Cu 0.01	100.00	3.67	or 0.6 di 2.7 cm 0.0 ab 7.3 ky 21.4 ap 0.7 an 41.4 tr 1.7 nf 17.0	Cc	S. L. Penfield	Am. Jour. Sci. 1886, 3, 32, 230. Mass anal. calc. by Farrington
117. Heredia	33.10	1.25	16.97	20.39	1.19	0.83	0.04	24.59	1.51	....	....	....	P₂O₅ 0.30	99.87	....	ab 6.8 di 4.8 nf 26.1 ky 15.3 ol 45.9	Ccb	I. Domeyko	Ann. de la Universidad de Chile 1859, 16, 335–339. Mass anal. calc. by Wadsworth
118. Cabezzo de Mayo	29.29	0.51	5.24	28.00	0.09	0.35	tr	13.66	1.37	....	Fe S 20.57	....	Chromite 0.02	100.00	....	ab 2.6 di 0.4 cm 0.0 ky 8.0 tr 20.6 ol 50.4 nf 15.0	Cw	S. Meunier	Thèse Faculté des Sciences de Paris, 1869, 9. Mass anal. calc. by Farrington

PERFEMIC, DOSILICIC, PERPOLIC, PEROLIC, PERMIRLIC, PERMIRIC, DOMAGNESIC, ORNANSOSE.

Name	Si O₂	Al₂ O₃	Fe O	Mg O	Ca O	Na₂ O	K₂ O	Fe	Ni	Co	S	P	Miscellaneous	Sum	Sp. gr.	Norm	Brezina's Symbol	Analyst	Reference
119. Shytal	32.05	2.54	23.88	22.90	1.12	1.50	0.67	10.38	1.63	....	0.78	0.05	NiO 0.86 Cu 0.11	98.47	3.55	lc 3.1 ns 0.7 tr 2.1 ne 5.1 di 1.8 sc 0.4 cm 1.2 nf 12.1	Cib	T. Hein	Sitzber. Wien Akad. 1866, 54, 2, 558–561
120. Ornans	31.23	4.32	24.71	24.40	2.27	0.55	....	4.12	1.85	....	2.69	....	Chromite 0.40 NiO 2.88 MnO tr	99.42	3.60	ab 2.6 di 69.4 cm 0.4 an 0.2 mn 0.8 tr 7.4 mo 2.3 nf 6.0	Cco	F. Pisani	Comptes Rendus, 1868, 67, 663–665
121. Cold Bokkeveld	30.80	2.05	29.94	22.20	1.70	1.23	....	2.50	....	tr	3.38	....	Cr₂O₃ 0.76 CuO 0.03 NiO 1.30 C 1.67 MnO 0.07 Bit. 0.25	98.78	2.69	ne 5.7 di 1.1 cm 1.1 an 0.3 ol 72.5 tr 9.2 cm 1.1 nf 2.5	K	E. P. Harris	Sitzber. Wien Akad. 1859, 35, 512
122. Mount Vernon	22.95	0.27	13.20	26.68	....	....	....	27.66	4.71	0.32	Fe S 0.69	Fe₃P 1.95	Fe₂O₃ 0.11 Cu 0.03 Chromite 1.00 Graphite 0.09 NiO 0.13 Al 0.12 MnO 0.09	100.00	....	C 0.3 ol 58.0 mtd 0.2 mo 4.1 cm 1.0 tr 0.7 sr 2.0 nf 32.8	P	Wirt Tassin	Proc. U. S. Nat. Mus. 1905, 28, 213–217. Mass anal. calc. by Farrington

PERFEMIC, DOMETALLIC, PERPOLIC, PERPYRIC, PERMIRIC, PERMIRIC, DOMAGNESIC, STEINBACHOSE

Name	Si O₂	Al₂ O₃	Fe O	Mg O	Ca O	Na₂ O	K₂ O	Fe	Ni	Co	S	P	Miscellaneous	Sum	Sp. gr.	Norm	Brezina's Symbol	Analyst	Reference
123. Steinbach	27.47	0.68	3.49	8.48	0.70	0.48	0.48	45.71	4.95	0.12	Fe S 7.22	0.07	Chromite 0.32 Mn O 0.16 Schreibersite 0.15	100.00	....	Q 8.7 ns 1.2 cm 0.3 ab 3.7 di 2.8 sc 0.8 ky 25.8 tr 7.22 nf 50.8	S	Winkler	Nova Acta. der K. Leop. Carol. deutsch Akad. 1878, 40. Mass anal. calc. by Farrington

PERFEMIC, DOMETALLIC, PERPOLIC, PEROLIC, DOPYRIC, PERMIRLIC, PERMIRIC, DOMAGNESIC, MINCIOSE.

Name	Si O₂	Al₂ O₃	Fe O	Mg O	Ca O	Na₂ O	K₂ O	Fe	Ni	Co	S	P	Miscellaneous	Sum	Sp. gr.	Norm	Brezina's Symbol	Analyst	Reference
124. Mincy	20.64	3.55	8.88	8.08	2.71	....	....	49.18	5.73	0.16	Fe S 0.99	0.08		100.00	4.84	an 0.7 di 3.0 tr 1.0 ky 21.5 sc 0.6 ol 9.0 nf 55.1	M	J. E. Whitfield	Am. Jour. Sci. 1887, 3, 34, 468–460. Mass anal. calc. by Farrington

PERFEMIC, DOMETALLIC, PERPOLIC, PEROLIC, PERMIRLIC, PERMIRIC, PERMAGNESIC, MARJALAHTOSE.

Name	Si O₂	Al₂ O₃	Fe O	Mg O	Ca O	Na₂ O	K₂ O	Fe	Ni	Co	S	P	Miscellaneous	Sum	Sp. gr.	Norm	Brezina's Symbol	Analyst	Reference
125. Marjalahti	8.07	....	2.38	9.47	....	0.04	0.01	73.95	5.71	0.34	....	....	Cr O 0.03	100.00	....	ol 20.0 nf 80.0	P	L. H. Borgström	Die Met. von Hvittis u. Marjalahti, Helsingfors 1903, 57. Mass anal. calc. by Farrington

ADDITIONAL ANALYSES OF IRON METEORITES

The following analyses of iron meteorites have been made since the writer's compilation (Pubs. Field Museum Geol. Ser. 1907, 3, 59-110) or were overlooked in making that compilation.

COARSE OCTAHEDRITES

Name	Fe	Ni	Co	Cu	Cr	P	S	C	Si	Cl	Insol	Miscellaneous	Total	Sp. gr.	Analyst	Reference
Bohumilitz	90.77	7.72	1.22										99.71		O. Koestler	1891, A. N. H. Wien, 6, 144
Cosby	89.72	10.12	0.42			0.11	tr						100.37		R. v. Reichenbach	1861, Pogg. Ann. 94, 250
Nuleri	93.57	5.79	0.41	tr		0.13	tr	0.01		tr			100.00	7.79	E. S. Simpson	1907, Bull. Geol. Survey, W. Australia, 26, 24-26
Wichita	90.77	8.34	0.26	0.02		0.14	0.02						99.88		J. W. Mallet	1884, A. J. S. 3, 28, 287
Wichita	91.39	7.91	0.40	tr								Sn 0.04, FeO +SiO₂ 1.32 Graphite 0.10	99.70		Cohen and Weinschenk	1891, A. N. H. Wien, 6, 153
Wichita	92.37	6.74	0.59	0.03		0.03							99.76		Manteuffel	1892, A. N. H. Wien, 7, 155

MEDIUM OCTAHEDRITES

Name	Fe	Ni	Co	Cu	Cr	P	S	C	Si	Cl	Insol	Miscellaneous	Total	Sp. gr.	Analyst	Reference
Ivanpah	91.12	6.92	1.73										99.77		O. Koestler	1891, A. N. H. Wien, 6, 145
Ivanpah	92.68	7.43	0.66	0.01		0.03							100.81		Manteuffel	1892, A. N. H. Wien, 6, 149
Inca	90.73	8.20	0.22		0.35	0.23	tr						99.97	7.64	Halbach	1907, Neues Jahrb. Festband. 230
Ilimäe	91.53	7.14	0.41	tr		0.44		0.24					99.52		C. Ludwig	1871, Sitzb. Wien Akad. 194
Joe Wright	91.67	7.53	0.99			tr	tr						100.19		Cohen and Weinschenk	1891, A. N. H. Wien, 6, 158
Rancho de la Pila	91.78	8.35	0.01			tr		tr					100.14		Janke	1884, Beitr. Abh. natur. Ver. Bremen, 8, 517
Tanokami	90.11	8.56	0.62			0.43							99.95	7.60	Kodera	1906, Beitr. z. Min. Japan, 2, 30-52
Williamstown	91.54	7.26	0.52	0.03	0.05	0.12	0.17	tr	tr				99.69	8.10	W. Tassin	1908, A. J. S. 4, 25, 49-50

FINE OCTAHEDRITES

Name	Fe	Ni	Co	Cu	Cr	P	S	C	Si	Cl	Insol	Miscellaneous	Total	Sp. gr.	Analyst	Reference
Muonionalusta	91.10	8.02	0.69	0.01	0.01	0.05							99.88	7.89	R. Mauzelius	1909, Bull. Geol. Inst. Univ. Upsala, 9, 236

BRECCIATED OCTAHEDRITES

Name	Fe	Ni	Co	Cu	Cr	P	S	C	Si	Cl	Insol	Miscellaneous	Total	Sp. gr.	Analyst	Reference
Ainsworth	92.22	6.49	0.42	0.01	0.01	0.28	0.07	0.09	0.05				99.64	7.85	W. Tassin	1908, A. J. S. 4, 25, 107

ATAXITES

Name	Fe	Ni	Co	Cu	Cr	P	S	C	Si	Cl	Insol	Miscellaneous	Total	Sp. gr.	Analyst	Reference
Guffey	88.69	10.55		0.02		0.02	0.02	0.02				Mn tr	99.87	7.94	Booth, Garrett and Blair	1909, Am. Mus. Jour. 9, 243
Weaver	81.81	16.63	1.18			tr	tr					Mn tr	99.62	7.99	F. Hawley	1910, Mineralogy of Arizona, 22
Weaver	79.60	18.80	1.60			tr	tr						100.00	7.98	W. B. Alexander	Same

FIELD COLUMBIAN MUSEUM

PUBLICATION 77

GEOLOGICAL SERIES   VOL. II, NO. 2

CATALOGUE

OF THE

COLLECTION OF METEORITES

MAY 1, 1903

BY

OLIVER CUMMINGS FARRINGTON, PH. D.

Curator, Department of Geology

· CHICAGO, U. S. A.

May 15, 1903

CATALOGUE OF THE COLLECTION OF METEORITES, MAY 1, 1903.

BY

OLIVER CUMMINGS FARRINGTON.

INTRODUCTION.

Since the first catalogue of the meteorite collection was issued * over seventy falls have been added and a number of other changes made.

It has, therefore, seemed desirable to prepare a second catalogue, not only to record the growth of the collection, but also to contribute new data to the subject of meteorites in general.

The following table shows the additions to the collection:

	July 15, 1895.	May 1, 1903.
Number of falls represented	179	251
Total weight	2,099 kgs. (4,627 lbs.)	2,289 kgs. (5,048 lbs.)
Number of casts	47	61
Number of micro-sections	5	26

The weight of the collection in 1895, as given above, has been changed from that stated in the first catalogue by the deduction of the Santa Catharina irons, formerly regarded as meteoric.

The following falls have an especially large representation in the collection: Bluff, Brenham, Cañon Diablo, Colfax, Crab Orchard, Farmington, Forest City, Hopewell Mounds, Indian Valley, Kenton County, Long Island, Saline, and Toluca. The weights of these falls possessed by the Museum, as compared with the original weights, are as follows:

	Original Weight.	Weight in Museum Collection.
Bluff	146 kgs.	13.9 kgs.
Brenham	900 "	445 "
Cañon Diablo	{ Indeterminate. { At least 4,000 "	689.8 "
Colfax	2.4 "	1.2 "
Crab Orchard	43 "	12.2 "
Farmington	84 "	23.5 "
Forest City	{ Indeterminate. { At least 122 "	16.5 "
Hopewell Mounds	At least 0.15 "	0.14 "
Indian Valley	14.2 "	8 "
Kenton County	163 "	56.3 "
Long Island	At least 564 "	528 "
Saline	31.13 "	20.5 "
Toluca	{ Indeterminate. { At least 1,000 "	177 "

* Handbook and Catalogue of the Meteorite Collection, Field Col. Mus. Pub. 3, Geol. Ser Vol. I, No. 1, Aug. 1895.
Field Col. Mus., Geol. Ser., Vol. II., No. 2.

The names by which the falls are designated in the catalogue have been chosen with considerable care, and in a few instances depart from previous usage. According to the writer's view, the name by which a meteorite is distinguished should be that of the nearest named locality. The desirability of this in enabling the place of fall or find quickly to be located at any time, without looking up the literature of the fall, can hardly be gainsaid. The exact point of fall or find is a datum of importance, and one whose significance is becoming more and more appreciated. If this datum were in all cases expressed in the name of the meteorite, a vast amount of vexatious labor would be saved to students of meteorites. In the case of the meteorites of the United States, the custom, often traceable to foreign authorities, of calling a meteorite by the name of the county in which it fell or was found is especially to be deprecated. Counties in this country often cover an area of hundreds of square miles. In process of time more than one fall or find is likely to occur within one county, causing a duplication of names if the plan of naming a fall from the county is consistently followed. Another change likely to occur is a subdivision of one county into two or more, thus throwing the place of fall or find out of the county for which it was named. An instance of this is afforded by Auburn, which was originally described from Macon County, but is now in Lee County. Similarly, Ruff's Mountain was formerly in Lexington County, but is now in Newberry County. It is true that in some instances the information regarding a place or find is so indefinite that it can only be designated as within a county, but wherever possible a more definite location should be determined, and the meteorite known by this name, even at the risk of disturbing established usage. That so many American meteorites have continued to be designated by the names of counties has probably been in part due to a lack of appreciation by foreign authorities of the amount of territory included in an American county.

The arrangement of names adopted in the accompanying catalogue is a purely alphabetical one, this being, according to the writer's experience, the most convenient for reference. In the classification of individual meteorites, the systems of Tschermak and Brezina have been followed, with such modifications and new determinations as are due to Wülfing, Cohen, Berwerth, Merrill, and others. The German terms have been put in English form for this catalogue, but the group abbreviations have been retained as in the original. Some falls hitherto without classification have been examined with a view to designation, and the following new determinations made:

Colfax, medium octahedrite, Om.

Hopper (Henry County), medium octahedrite, Om.

Saline, crystalline spherical chondrite, Cck.

Study of the specimens of some falls in the collection has seemed to indicate that their present accepted classification is incorrect, and accordingly the following changes are recommended:

Baratta—Intermediate chondrite, Ci, instead of black chondrite, Cs.

Bridgewater—Medium octahedrite, Om, instead of fine octahedrite, Of.

Indian Valley — Hexahedrite, H, instead of brecciated hexahedrite, Hb.

Lançon—Brecciated gray chondrite, Cgb, instead of white chondrite, Cw.

Tysnes—Brecciated intermediate chondrite, Cib, instead of brecciated gray chondrite.

The additions received to the collection since its inception in 1894 have been obtained by gift, exchange, and purchase. The accessions by gift include a large Cañon Diablo individual, from the Ed. E. Ayer Pioneer Hose Co.; a small individual of the same fall, from George Bell; a small individual of the Estherville fall, from A. E. J. Svege; and sections of the Allegan, Oakley, and St. Genevieve County meteorites, from Prof. H. A. Ward. General G. Murray Guion has kindly continued the loan of the Seneca Falls specimen.

Exchanges conducted with a number of institutions and individuals call for acknowledgment of courtesies from the following:

S. C. H. Bailey, Julius Böhm, Dr. A. Brezina, the British Museum through Prof. L. Fletcher, Prof. E. Cohen, E. E. Howell, Institute of Mines, St. Petersburg, through Prof. Melnikoff, Max Neumann, B. Stürtz, the United States National Museum through Dr. George P. Merrill, the Vienna Museum through Prof. F. Berwerth, Prof. H. A. Ward, Ward's Natural Science Establishment, and the University of Wisconsin through Prof. W. H. Hobbs.

The plates accompanying the catalogue illustrate for the most part specimens obtained subsequent to the publication of the first catalogue. Each will be found described under the corresponding fall in the body of the work.

The clerical work involved in the preparation of the catalogue has been performed with care and accuracy by Mr. Louis V. Kenkel of the Department of Geology.

CATALOGUE OF THE COLLECTION.

No.	Date of Fall or Find.	NAME AND DESCRIPTION.	Weight in grams.
1	Found 1881.	**Admire,** Lyons Co., K a n s a s. Iron-stone. Brecciated pallasite, Pb. Two polished sections showing nickel-iron and chrysolite. Structure brecciated. Much oxidized. Cat. No., 557.	59.4
2	Fell 1814, Sept. 5, Noon.	**Agen,** Dep. Lot-et-Garonne, F r a n c e. Stone. Veined intermediate chondrite, Cia. Fragment from interior, showing white chondri and metallic grains distributed through a darker ground mass. Cat. No., 235. Fragment, with crust. Interior, light-gray with darker spots having the appearance of brecciation rather than of veining; crust, black. Cat. No., 526.	0.5 85
3	Fell 1806, Mar. 15, 5 P. M.	**Alais,** Dep. Gard, F r a n c e. Stone. Carbonaceous chondrite, K. Coarse, brown-black powder resembling an earthy coal. Very friable. Cat. No., 221. Interior fragment like previous specimen. Cat. No., 222.	1 0.4
4	Fell 1873.	**Aleppo,** S y r i a. Stone. Brecciated white chondrite, Cwb. Fragment, with crust. Crust, black, 2 mm. in thickness. Interior, light-gray. Friable. No distinct chondri are visible. Cat. No., 544.	1.45
5	Fell 1883, Feb. 16, 3 P. M.	**Alfianello,** Brescia, I t a l y. Stone. Intermediate chondrite, Ci. Fragment, with crust. The latter is about .4 mm. in thickness, and of a dirty black color. The interior of the stone is ash-gray, fine-grained, and contains metallic grains, with some coarse nodules of the same. Cat. No., 332. Interior fragment, ash-gray, with brown spots, due to the oxidation of the metallic particles. Several of the latter are quite large, and rounded as if previously fused. Cat. No., 333. Large fragment, with crust. Characters like those of previous specimen. Cat. No., 334.	24 134.5 300
6	Found 1887.	**Algoma,** Kewaunee Co., W i s c o n s i n. Iron Medium octahedrite, Om. Etched fragment, with crust, well marked Widmanstätten figures, with kamacite uniformly bordered by taenite. Cat. No., 529.	6.5
7	Fell 1899, July 10, 8 A. M.	**Allegan,** Allegan Co., M i c h i g a n. Stone. Spherical chondrite, Cc. Fragment, friable, with few metallic grains. Chondri readily separable. Cat. No., 498. Fragment, with crust. Crust, dirty black, porous. Gift of Prof. H. A. Ward. Cat. No., 500.	86 56

No.	Date of Fall or Find.	NAME AND DESCRIPTION.	Weight in grams.
8	Found 1898.	**Arispe,** Sonora, M e x i c o. Iron. Coarsest octahedrite, Ogg. Etched slab, 12.5 x 10.5 cm., with crust. Shows coarse Widmanstätten figures, and troilite distributed in vein-like manner. There is apparent faulting of the Widmanstätten figures on either side of the vein. Cat. No., 554.	810
9	Found 1894.	**Arlington,** Sibley Co., M i n n e s o t a. Iron. Medium octahedrite, Om. Etched fragment, with crust. Shows typical octahedral figures, with the kamacite uniformly bordered by taenite. See Plate XXXII. Cat. No., 459.	70
10	Found 1867.	**Auburn,** Lee Co., A l a b a m a. This locality is usually given as Auburn, Macon Co. This was correct up to 1866, but since that year Auburn has been in Lee Co. Iron. Hexahedrite, H. Sawed fragment showing crust on all surfaces but one. Cat. No., 92.	5
11	Found 1890.	**Augustinowka,** Gov. Ekaterinoslaw, R u s s i a. Iron. Medium octahedrite, Om. Sawed block, with crust. Etched. Shows uniform Widmanstätten figures and veins and inclusions of troilite. Cat. No., 508. Fragment of crust. Much oxidized, and colored green by nickel salts. Cat. No., 507.	217 60
		Aumale, see Senhadja.	
12	Fell 1842, June 3.	**Aumières,** Dep. Lozère, F r a n c e. Stone. Veined white chondrite, Cwa. Fragment, with crust. Interior light-gray, with minute, shining metallic grains. Rather friable. Individual chondri not discernible. Cat. No., 491.	8.5
13	Fell 1858, Dec. 9, 7:30 A. M.	**Aussun** (Clarac), Dep. Haute Garonne, F r a n c e. Stone. Spherical chondrite, Cc. Fragment from interior. Light-gray, with rusty-iron grains. Compact. Delicate veins penetrate the mass. Cat. No., 271. Fragment from the interior, with polished surface. The latter shows large chondri of a chrysolite-like mineral, embedded in a ground mass made up chiefly of small white chondri and grains of nickel-iron. Cat. No., 272.	13 22
14	Known 1842.	**Babb's Mill,** Green Co., T e n n e s s e e. Iron. Nickel-rich ataxite. Morradal group. Polished slab. Cat. No., 109.	70
15	Fell 1814, Feb. 15, Noon.	**Bachmut** (Alexejewka), Gov. Ekaterinoslaw, R u s s i a. Stone. White chondrite, Cw. Fragment, with crust and polished surface. Crust, dull-black. Interior, light-gray, with a few rusty grains. Cat. No., 234.	12

No.	Date of Fall or Find.	NAME AND DESCRIPTION.	Weight in grams.
16	Found 1871.	**Bacubirito** (Ranchito), Sinaloa, M e x i c o. Iron. Finest octahedrite, Off. Thick, etched slab with crust. Very narrow Widmanstätten figures are plainly defined. There are some angular partings and two small troilite inclusions. See Plate XXXV. Cat. No., 537.	352
17	Found 1893.	**Ballinoo,** A u s t r a l i a. Iron. Finest octahedrite, Off. Polished and etched slab with crust. Crust surface smooth and little oxidized, indicating a recent fall. Etched surface shows no figures. Cat. No., 456.	157
18	Fell 1871, Dec. 10, 1:30 P. M.	**Bandong,** Goemoroeh, J a v a. Stone. Brecciated white chondrite, Cwb. Two fragments from interior. Grayish-brown, with metallic particles. Cat. No., 304.	1.5
19	Fell 1845, May, 5:15 P. M.	**Baratta,** Deniliquin, A u s t r a l i a. Stone. Intermediate chondrite, Ci. Full-sized section, 10 x 15 cm. one surface polished, with crust. Polished surface is thickly sprinkled with gray and white chondri of various sizes. Metallic grains small in size and amount. Many chondri show a rim of metal. Crust thin, black, somewhat rusty. See Plate XXXI. Cat. No., 539.	295
20	Fell 1790, July 24, 9 P. M.	**Barbotan,** Gascony, F r a n c e. Stone. Gray chondrite, Cg. Fragment from interior, one surface polished, showing numerous, minute, metallic grains. Cat. No., 213. Fragment with crust, showing pitted surface. Much discolored by age. Cat. No., 214.	6 32
21	Fell 1892, Aug. 29, 4 P. M.	**Bath,** Brown Co., S o u t h D a k o t a. Stone. Brecciated spherical chondrite, Ccb. Irregular fragment, with crust and polished surface. The crust surface is indented with broad, shallow pits. Crust, dull-black, papillated, not more than .3 mm. in thickness. Interior, grayish-brown, of fine-granular structure, containing minute metallic grains. A portion shows "slickensided" surface. Cat. No., 351.	1,276
22	Fell 1902, Nov. 15, 6:45 P. M.	**Bath Furnace,** Bath Co., K e n t u c k y. Stone. Intermediate chondrite, Ci. Fragment, with crust and three sawed surfaces. Interior gray, with rusty spots and thickly sprinkled metallic grains. The latter are often irregular in shape and show bronze-yellow inclusions of troilite. The texture of the stone is quite firm. The presence of chondri is indicated by darker and lighter spots. Crust, black and uniform. Shows typical pittings. See Plate XXXII. Cat. No., 555.	366

No.	Date of Fall or Find.	NAME AND DESCRIPTION.	Weight in grams.
23	Described 1887.	**Beaconsfield,** Victoria Colony, A u s t r a l i a. Iron. Coarse octahedrite, Og. Etched section showing distinct Widmanstätten figures. Contains two troilite nodules each rimmed by schreibersite. Cat. No., 499.	413
24	Found 1866.	**Bear Creek,** Jefferson Co., C o l o r a d o. Iron. Fine octahedrite, Of. Fragment showing crust. Octahedral cleavage well displayed. Cat. No., 88. Thin slab, etched. Well marked Widmanstätten figures, the plates of taenite being very distinct. Cat. No., 89.	43.5 27
25	Fell 1893, May 26, 3:30 P. M.	**Beaver Creek,** West Kootenai District, B r i t i s h C o l u m b i a. Stone. Crystalline spherical chondrite, Cck. Fragment, with crust. Interior, dark-gray, made up of small, glassy chondri, and fine metallic grains. Cat. No., 352. Like previous specimen. Crust, dull-black, about .3 mm. thick. Cat. No., 353.	7 19
26	Found 1784.	**Bemdegó,** Bahia, B r a z i l. Iron. Coarse octahedrite, Og. Scalings from crust, showing magnetite. Cat. No., 3. Slab, 8 x 17 cm., showing natural and etched surfaces. The etched surface exhibits a coarsely crystalline structure with imperfect Widmanstätten figures. Cat. No., 5. Etched slab, 9 x 23 cm., showing imperfect Widmanstätten figures, elongated nodules of troilite and a group of schreibersite inclusions. Cat. No., 6.	1,132 855
27	Fell 1798, Dec. 19, 8 P. M.	**Benares,** Bengal, I n d i a. Stone. Spherical chondrite, Cc. Powder, showing crust and individual chondri. Cat. No., 216. Fragment, with crust. One surface polished, showing scattered metallic grains. Cat. No., 217.	.71 1
28	Fell 1811, July 8, 8 P. M.	**Berlanguillas,** Burgos, S p a i n. Stone. Veined intermediate chondrite, Cia. Polished fragment. Texture firm. Abundant metallic grains. Cat. No., 492.	5.5
29	Fell 1843, March 25.	**Bishopville,** Sumter Co., S o u t h C a r o l i n a. Stone. Veined chladnite, Chla. Fragment from interior. Light-gray, with white nodules of the chladnite of Shepard. Rusty brown spots show the presence of metallic grains. Cat. No., 251. Like previous specimen, but showing vitreous crust. Cat. No., 252. Fragment of chladnite. Cat. No., 253.	1 2 5

No.	Date of Fall or Find.	NAME AND DESCRIPTION.	Weight in grams.
30	Found 1802.	**Bitburg** (Albacher Mühle), R h e n i s h P r u s s i a. Iron-stone. Pallasite P. Polished slab showing large pores and slag-like surface, due to its having been passed through a furnace. Cat., No., 30.	22
		Fragment, three sides polished. The natural surface appears to be altering to limonite. Cat. No., 31.	72
31	Fell 1887, Jan. 1.	**Bjelokrynitschie,** Volhynia, R u s s i a. Stone. Brecciated intermediate chondrite, Cib. Polished fragment with crust. Mottled gray and brown, groundmass with numerous metallic grains of irregular size. Compact in texture. Crust black, firmly adhering. Cat. No., 484.	9
32	Fell 1899, March 12, 10:30 P. M.	**Bjurböle,** near Borgo, F i n l a n d. Stone. Spherical chondrite, Cc. Fragment with crust. Stone friable, and chondri of the size of peas and smaller, are easily separable. There are angular inclusions of a homogeneous, fine-grained rock also. The crust is black, coarse, and shows in places lines of flow. Metallic grains scarce. See Plate XXXII. Cat. No., 522.	310
33	Found 1878.	**Bluff,** Fayette Co., T e x a s. Stone. Brecciated crystalline chondrite, Ckb. About one-tenth the original mass, showing crust and polished surface 25 x 35 cm. The crust surface is somewhat decomposed, but shows the characteristic pittings. The polished surface shows the dark-green color of the stone, with its fine texture and scattered metallic grains. Cat. No., 173.	10,985
		Thin slab 31 x 38 cm., exhibiting the black veins peculiar to this meteorite. Cat. No., 174.	2,934
34	Found 1829.	**Bohumilitz,** Bohemia, A u s t r i a. Iron. Coarse octahedrite, Og. Etched slab, 6.5 x 9 cm. with crust. Shows typical swollen octahedral figures. Cat. No., 487.	275
35	Fell 1852, Oct. 13, 3 P. M.	**Borkut,** Marmaros, H u n g a r y. Stone. Spherical chondrite, Cc. Fragments showing individual chondri, spheroidal, dark-green in color. Cat. No., 262.	.12
36	Fell 1847, July 14, 3:45 A. M.	**Braunau,** Hauptmannsdorf, B o h e m i a. Iron. Hexahedrite, H. Sawed block showing natural surface with pits. The luster of the natural surface is like that of blued steel. Cat. No., 55.	47
		Brazos River, see Wichita Co.	

No.	Date of Fall or Find.	NAME AND DESCRIPTION.	Weight in grams.
37	Fell 1855, May 13, 5 P. M.	**Bremervörde,** (Gnarrenburg), Prov. Hanover, Germany. Stone. Brecciated spherical chondrite, Ccb. Fragment from interior. Cat. No. 265. Fragment with crust, and polished surface. The polished surface is gray, mottled with white chondri and irregular metallic grains, some of them large. Crust black, thin. Cat. No., 438.	.5 72
38	Found 1882.	**Brenham,** Brenham Township, Kiowa Co., Kansas. Iron-stone. Pallasite, P. One-half of a complete individual, one surface polished. Composed chiefly of iron, with chrysolite filling the sponge-like pores. Cat. No., 195. Thin slab, polished. The central portion for a width of about 5 cm. is solid metal, but on either side the mass is porous, the pores being filled with chrysolite. Cat. No., 196. Full-sized slab, polished, showing a sponge-like mass of iron, with chrysolite filling the cavities. Cat. No., 197. Similar to No. 197, but thicker. Some of the chrysolite nodules are beautifully transparent and highly refracting. Cat. No., 198. Smaller piece, similar to No. 197. Cat. No., 199. 466 pound mass, entire. The form is flattened, somewhat heart-shaped. The surface is covered with pittings, and considerably oxidized. The grains of chrysolite are readily discernible over the surface. Cat. No., 200. Full-sized slab, showing structure like No. 197. Cat. No., 201. 18 pound mass, entire. Form, hemispheroidal, the surface covered with pittings. Structure porous, pores filled with chrysolite. Cat. No., 202. Section of complete individual, showing natural and polished surface. The structure is like that of No. 197. Cat. No., 203. 345 pound mass, entire. This is almost wholly iron. Form, kidney or arch-shaped, with a projection extending from the concavity of the arch. Cat. No., 204. 36 pound mass, entire. Spheroidal in form, surface covered with pittings. Entirely metallic. Cat. No., 205. 40 pound mass, entire. Form, cylindrical, with one projecting point. Surface pitted. Composed almost wholly of iron. but occasional grains of chrysolite are visible. Cat. No., 206. Etched section with crust. Shows typical octahedral figures. Cat. No., 432. Section showing iron and chrysolite. Cat. No., 450. Like No. 450. Cat. No., 436.	2,061 1,248 2,048 8,117 227 218,847 5,895 8,619 8,490 155,473 16,091 17,687 160 165 36

No.	Date of Fall or Find.	NAME AND DESCRIPTION.	Weight in grams.
39	Described 1890.	**Bridgewater,** Burke Co., North Carolina. Iron. Medium octahedrite, Om. Thin slab with natural and etched surfaces. Exhibits well marked Widmanstätten figures. Cat. No., 137.	19
40	Found before 1819.	**Burlington,** Otsego Co., New York. Iron. Medium octahedrite, Om. Triangular slab, etched on one surface. Shows very delicate Widmanstätten figures. Cat. No., 36.	32
41	Found before 1874.	**Butler,** Bates Co., Missouri. Iron. Finest octahedrite, Off. Etched slab. Widmanstätten figures very distinct. The plates of the latter seen with a lens appear to be made up of a number of smaller ones, which anastomose. There are also comb-like markings, made up of innumerable fine lines. Contains nodule of troilite. Cat. No., 96.	71.5
42	Fell 1861, May 12, about noon.	**Butsura,** Goruckpur, India. Stone. Intermediate chondrite, Ci. Fragments with crust and polished surfaces. Nickel-iron is present in large amount. There are also chondri 1 mm. in diameter, of a chrysolite-like mineral. The ground-mass is greenish black and structureless. Cat. No., 277.	1.85
		Cabarrus County, see Flows. **Caille,** see La Caille.	
43	Found 1891.	**Cañon Diablo,** Coconino Co., Arizona. Iron. Coarse octahedrite, Og. Complete individual, apparently scaled off from a larger mass. Shows the smooth surface and characteristic pits of this iron. Cat. No., 141.	844
		Complete individual. Besides the shallow pits shown in the figure the mass is indented by deeper cylindrical ones, 3 to 4 cm. in depth. Cat. No., 143.	460,304
		Full-sized section, 20 x 29 cm., with polished and etched surfaces. The Widmanstätten figures are very coarse and arranged in approximately parallel bands. Large nodules of troilite and flakes of schreibersite are scattered through the mass. Cat. No., 144.	2,934
		Small, complete individual, like No. 141. Cat. No., 145.	165
		Complete individual. It contains a natural perforation about 3 cm. in its smallest diameter. Cat. No., 146.	120,657
		Full-sized slab, 19 x 30 cm., showing polished and etched surfaces like No. 144. Cat. No., 147.	4,309
		Nearly complete individual showing deep and shallow pits. One etched surface, 19 x 21 cm., exhibits nodules of troilite and indications of crystalline structure. Cat. No., 148.	23,590

No.	Date of Fall or Find.	NAME AND DESCRIPTION.	Weight in grams.
		Complete individual, sub-cylindrical in form. Exhibits the characteristic pittings. Cat. No., 149.	90,898
		Thick slab, 20 x 27 cm., polished, showing nodules of troilite. Cat. No., 150.	26,047
		Hemispheroidal mass. One polished surface, 20 x 25 cm., shows troilite nodules Cat. No., 151.	24,489
		Six small fragments with natural surface. Cat. No., 152.	123
		Complete individual, showing pits and natural surface. Apparently scaled from a larger mass. Cat. No., 373.	620
		Complete individual. Cat. No., 455.	60
		Gift of George Bell.	
		Complete individual, deeply pitted. Cat. No., 497.	34,800
		Gift of Ed. E. Ayer Pioneer Hose Company.	
44	Found 1793.	**Cape of Good Hope,** A f r i c a. Iron. Ataxite, with hexahedral streaks. Polished slab of brilliant nickel-white color. Cat. No., 29.	27
45	Found 1887.	**Carlton,** Hamilton Co., T e x a s. Iron. Fine octahedrite, Of. Full-sized, thin slab, 22 x 30 cm., showing polished and etched surfaces. The Widmanstätten figures appear as beautifully distinct and delicate lines running parallel in two directions throughout the mass. Troilite is distributed in radiating veins. Cat No., 131.	3,406
46	Found 1840.	**Carthage,** Smith Co., T e n n e s s e e. Iron. Medium octahedrite, Om. Thick slab showing natural, polished and etched surfaces Coarse Widmanstätten figures are dimly outlined on the latter. The lines of taenite are very delicate. Cat. No., 51.	50
47	Recognized 1867.	**Casas Grandes,** Chihuahua, M e x i c o. Iron. Medium octahedrite, Om. Etched section 12 x 13 cm., with crust. Shows typical Widmanstätten figures, the bands of kamacite being long and well-defined. Little taenite is to be seen and little troilite. Cat. No., 524.	608
48	Fell 1874, May 14, 2:30 P. M.	**Castalia,** Nash Co., N o r t h C a r o l i n a. Stone. Gray chondrite, Cg. Fragment with crust; the latter dull-black and scoriaceous. The color of the stone is dark-gray, with no metallic grains visible. Cat. No., 307.	1.5
		Fragment from interior, showing occasional metallic grains. Cat. No., 308.	6.5
49	Fell 1838, June 6, Noon.	**Chandakapoor,** Berar, I n d i a. Stone. Brecciated intermediate chondrite, Cib. Three fragments from interior, two of them polished, showing a dark-gray stone, containing numerous rusty iron grains. Cat. No., 245.	3.5

No.	Date of Fall or Find.	NAME AND DESCRIPTION.	Weight in grams.
50	Fell 1812, Aug. 5, 2 A. M.	**Chantonnay,** Dep. Vendée, F r a n c e. Stone. Gray chondrite, Cg. Thin chip, highly polished. Almost black, with few metallic grains. Structure not discernible megascopically. Cat. No., 232. Polished fragment with crust. Cat. No., 443.	2 4
51	Known 1804.	**Charcas,** San Luis Potosi, M e x i c o. Iron. Medium octahedrite, Om. Etched section with crust. Well marked Widmanstätten figures. Cat. No., 33.	62
52	Fell 1835, Aug. 1, 2-3 P. M.	**Charlotte,** Dickson Co., T e n n e s s e e. Iron. Fine octahedrite, Of. Thin slab, one surface etched. Typical Widmanstätten figures. Cat. No., 40.	7
53	Fell 1810, Nov. 23, 1:30 P. M.	**Charsonville,** near Orleans, Dep. Loiret, F r a n c e. Stone. Veined gray chondrite, Cga. Fragment from interior. Light-gray, with rusty-brown spots, due to oxidation of the abundant metallic grains. Cat No., 229. Thin chip, polished. Like previous specimen, but traversed by a delicate black vein. Cat. No., 230.	22 2
54	Fell 1815, Oct. 3, 8 A. M.	**Chassigny,** Dep. Haute-Marne, F r a n c e. Stone. Chassignite, Cha. Fragments from the interior. Composed of light-yellow friable chrysolite specked with black. Cat. No., 546.	0.7
55	Fell 1841, June 12, 1:30 P. M.	**Chateau Renard,** Dep. Loiret, F r a n c e. Stone. Veined intermediate chondrite, Cia. Fragment from interior. Gray, compact, traversed by delicate black veins. Metallic grains small and bright. Cat. No., 249. Like previous specimen. Cat. No., 248. Like No. 249. Cat. No., 247.	57 7 5
56	Found before 1849.	**Chesterville,** Chester Co., S o u t h C a r o l i n a. Iron. Nickel-poor ataxite, Nedagolla group. Thin slab, etched. The etching brings out a network of irregular lines on the surface. Cat. No., 56.	6
57	Found 1873.	**Chulafinnee,** Cleburne Co., A l a b a m a. Iron. Medium octahedrite, Om. Thin slab, etched. Broad Widmanstätten figures are dimly outlined. Cat. No., 94.	29
58	Found 1860.	**Cleveland,** Bradley Co., T e n n e s s e e. Lea Iron. Probably the same as Dalton. Iron. Medium octahedrite, Om. Large, thin slab, showing crust, polished and etched surfaces. Typical Widmanstätten figures. Cat. No., 127.	234

No.	Date of Fall or Find.	NAME AND DESCRIPTION.	Weight in grams.

59	Known 1837.
	
	Found 1868.
		Coahuila, Mexico.
		   Iron. Hexahedrite, H. (Sancha Estate.) Thin, polished slab, a portion etched. Cat. No., 37.
		   Turnings. Cat. No., 90.
		   Butcher Iron. Large, thin slab, a portion etched. The latter shows a stippled surface intersected by numerous Neumann lines. Included nodules of troilite are also to be seen. Cat. No., 42.
		   Large segment, showing natural, polished and etched surfaces. Natural surface very smooth. Etched surface like that of previous specimen. Cat. No., 43.
			70
			10
			
			2,140
			
			3,402

60	Fell 1838, Oct. 13, 9 A. M.	**Cold Bokkeveld,** Cape of Good Hope, Africa. Stone. Carbonaceous chondrite, K. Fragment from interior. Dull-black with white specks. Resembles a piece of graphite. Cat. No., 246.	1

61	Found 1880.	**Colfax,** Colfax Township, near Ellenboro, Rutherford Co., North Carolina.
		   Iron. Medium octahedrite, Om. Spheroidal mass, showing natural and etched surfaces and fracture. The latter shows the iron to be highly crystalline, and to possess octahedral cleavage. Cat. No., 153.
		   Etched section, showing typical octahedral figures. See Plate XXXV. Cat. No., 156.
			1,130
			
			61

62	Fell 1890, Feb. 3, 1:30 P. M.	**Collescipoli,** Terni, Italy. Stone. Spherical chondrite, Cc. Fragment, with crust. Crust dull black, scoriaceous, nearly 2 mm. in thickness. Interior of stone light bluish-gray. Shows chondri and metallic grains. Cat. No., 356.	2.5
63	Known 1860.	**Coopertown,** Robertson Co., Tennessee. Iron. Medium octahedrite, Om. Thin slab, etched. The Widmanstätten figures are made up of broad plates, 5 mm. in thickness. Cat No., 83	82.5

64	Described 1840.	**Cosby's Creek,** Cocke Co., Tennessee.
		   Iron. Coarse octahedrite, Og. Several irregular fragments, all showing octahedral cleavage. Cat. No., 48.
		   Irregular fragment, cleavage structure prominent. Cat. No., 49.
		   Five fragments of about 20 grams each, cleaving in octahedrons, which are separated by bright plates of taenite. Cat. No., 53.
			35.5
			
			43
			
			114

65	Found 1881.	**Costilla Peak,** Taos Co., New Mexico. Iron. Medium octahedrite, Om. Nearly full-sized etched section, 11 x 20 cm. Shows typical octahedral figures. Cat. No., 502.	1,154

No.	Date of Fall or Find.	NAME AND DESCRIPTION.	Weight in grams.
66	Found 1887.	**Crab Orchard,** near Rockwood, Cumberland Co., Tennessee.	
		Iron-stone. Mesosiderite, M. From Mass No. 1. Thin slab showing natural and etched surfaces. The metallic grains are small and about evenly distributed, except for three large nodules, one of which, having a diameter of 1.5 cm., shows well-marked Widmanstätten figures. The metallic portion serves as a matrix to hold the siliceous grains. Cat. No., 184.	40.5
		Mass No. 2. Complete individual. The crust is reddish-brown and cracked in several places. No well-marked pits are seen. Cat. No., 185.	4,351
		Part of Mass No. 3. One-half of the original find, with natural and polished surface, 12 x 13 cm. General structure like No. 184, but the specimen shows a larger proportion of the non-metallic minerals, and these occur occasionally in large nodules. Cat. No., 186.	801
		From Mass No. 3. Large segment, 17 x 26 cm., showing natural and polished surfaces. Cat. No., 187.	6,151
		Segment from Mass No. 1, showing natural and polished surfaces. Structure like No. 184. Cat. No., 374.	806
		Irregular fragment, one surface polished. Metallic grains small and evenly distributed. Cat. No., 336.	80.4
67	Found 1854.	**Cranbourne,** Victoria, Australia.	
		Iron. Coarse octahedrite, Og. Irregular fragment, much decomposed. A portion altered to limonite. Silvery plates of taenite are numerous through the mass. Cat. No., 68.	34.2
		Cleaved fragment, octahedral structure prominent. Cat. No., 69.	4.5
		(Yarra Yarra River.)	
		Thin slab showing natural and etched surface. Crystalline structure is indicated on the etched surface, but no distinct Widmanstätten figures are shown. Cat. No., 70.	15
		Cross Timbers, see Red River.	
		There are three "Cross Timbers" in Texas, occurring in Denton, Harris, and Johnson counties respectively. None of these is near the locality 32° 7' N. and 95° 10' W., at which this meteorite is reported to have been found. It is true this locality is a long distance from the Red River as well, but this name has historic usage.	
68	Found 1877.	**Dalton,** Whitfield Co., Georgia.	
		Iron. Medium octahedrite, Om. Thin, etched section showing coarse, typical Widmanstätten figures and crust. Cat. No., 110.	81

No.	Date of Fall or Find.	NAME AND DESCRIPTION.	Weight in grams.
69	Fell 1878. Sept. 5.	**Dandapur,** Dist. Guorckpur, I n d i a. Stone. Veined intermediate chondrite, Cia. Fragment, with crust, one surface polished. Crust black, thin, minutely pitted. Interior light-gray, much spotted with rust. The outlines of light colored chondri can be discerned here and there. Metallic grains are rather numerous. Cat. No., 449.	27.5
70	Found 1846.	**Deep Springs,** Deep Springs Farm, Rockingham Co., N o r t h C a r o l i n a. Iron. Nickel-rich ataxite. Morradal group. Nearly complete etched section 7 x 17 cm. Exhibits no figures, but shows a few troilite inclusions. Cat. No., 453.	420
71	Known 1856.	**Denton County,** T e x a s. Iron. Medium octahedrite, Om. Thin, sawed fragment. Along an old fracture are numerous parallel grooves which are probably lines of decomposition. Cat. No., 74.	3
72	Found before 1780.	**Descubridora,** Catorce, San Luis Potosi, M e x i c o. Iron. Medium octahedrite, Om. Polished and etched slab. Widmanstätten figures in long parallel bands are crossed at intervals by others nearly at right angles. Cat. No., 2.	35
73	Fell 1860, July 14, 2:15 P. M.	**Dhurmsala,** Punjaub, I n d i a. Stone. Intermediate chondrite, Ci. Fragments from interior. Cat. No., 275. Fragment with crust, the latter black, shining and showing numerous pits. Interior light-gray, with rusty grains. Compact. Nodules of a bluish gray color, finer grained than the rest, are distributed through the mass. Cat. No., 276.	5 123
74	Found 1888.	**Doña Inez,** Atacama, C h i l e. Iron-stone. Mesosiderite, M. Thin slab, showing natural and etched surfaces. The stony matter, dark-brown in color, largely predominates. One nodule of iron about the size of a pea, shows delicate Widmanstätten figures. Cat. No., 191. Hemispheroidal mass, one surface polished. The peculiar cracked surface characteristic of these meteorites is well exhibited. Cat. No., 192. Complete individual, described by Howell as looking like "a lump of dried red mud cracked by shrinkage and covered with spots of green mould (nickel) in places." Cat. no., 193. Similar to No. 192, but larger. Cat. No., 194.	48.5 103 741 245

No.	Date of Fall or Find.	NAME AND DESCRIPTION.	Weight in grams.
75	Fell 1815, Feb. 18, Noon.	**Durala,** Punjaub, India. Stone. Veined intermediate chondrite, Cia. Sawed fragment with crust. Crust thick, dark and blebby. Interior gray, somewhat friable and showing coarse metallic grains. Cat. No., 559.	54
76	Found 1880.	**Eagle Station,** Carroll Co., Kentucky. Iron-stone. Pallasite, P. Sawed slab, showing natural surface deeply pitted, and polished surface. The iron matrix encloses fragments of chrysolite, some a centimeter in diameter, transparent and of brilliant luster. See Plate XXXIII. Cat. No., 180. Grains of chrysolite, separated from the iron, some coarse and some in a powdered state. Cat. No., 182.	106 6
77	Fell 1785, Feb. 19.	**Eichstädt,** Wittmess, Bavaria. Stone. Spherical chondrite, Cc. Fragment from interior, showing a gray, coarse ground-mass containing rusty iron grains. Cat. No., 212.	1
78'	Fell 1400? Recognized 1811.	**Elbogen,** Bohemia. Iron. Medium octahedrite, Om. Etched fragment showing Widmanstätten figures. Cat. No., 1.	2.5
79	Found 1893.	**El Capitan Mts.,** Lincoln Co., New Mexico. Iron. Medium octahedrite, Om. Etched section showing crust and typical octahedral figures. Cat. No., 430. **Ellenboro,** see Colfax.	214
80	Found 1854.	**Emmitsburg,** Frederick Co., Maryland. Iron. Medium octahedrite, Om. Etched slab with natural surface. Widmanstätten figures well brought out. Cat. No., 67.	28.5
81	Fell 1492, Nov. 16, 11:30 P. M.	**Ensisheim,** Upper Alsace, Germany. Stone. Crystalline chondrite, Ck. Fragment from interior. Dark-gray, fine grained, smooth and shining in portions. Cat. No., 207. Similar to previous specimen. Cat. No., 208.	22 4
82	Fell 1889, July.	**Ergheo,** Somaliland Peninsula, E. Africa. Stone. Crystalline chondrite, Ck. Polished section with crust. Polished surface brownish black, showing small metallic grains, rather thinly distributed. Texture firm, compact. Crust surface smooth and of reddish-brown color, indicating long exposure. Cat. No., 553.	192.5

No.	Date of Fall or Find.	NAME AND DESCRIPTION.	Weight in grams.
83	Fell 1812, Apr. 15, 4 P. M.	**Erxleben,** Saxony, P r u s s i a. Stone. Crystalline chondrite, Ck. Fragment from interior. Dark-gray, compact, made up of siliceous grains with a vitreous luster, and numerous fine, metallic grains. Cat. No., 231.	2.5
84	Fell 1879, May 10, 5 P. M.	**Estherville,** Emmet Co., I o w a. Iron-stone. Mesosiderite, M. Irregular fragment, much oxidized. Cat. No., 176.	75
		Full-sized slab, 18 x 31 cm., polished. Nickel-iron appears in large nodules, irregular flakes and a long, narrow vein, distributed through a greenish-black, structureless ground-mass. Cat. No., 177.	2,721
		Thirteen complete individuals, varying in size from a pea to a walnut. Surface like No. 175. Cat No., 178.	47
		Complete individual. Cat. No., 458.	1
		Gift of A. E. J. Svege.	
85	Fell 1890, June 25, 1 P. M.	**Farmington,** Farmington Township, Washington Co., K a n s a s. Stone. Black chondrite, Cs. Fragment from interior, having the appearance of a dolerite of dark-gray color and splintery fracture. Contains white radiated chondri. Bronze-yellow metallic grains are numerous. Cat. No., 342.	120
		Thin slab, polished, showing white and dark chondri, and various grains of nickeliferous iron. Cat. No., 344.	425
		Similar to No. 343. Cat. No., 345.	672
		Full-sized slab, 15 x 40 cm., polished. Similar to above specimen. Cat. No., 346.	1,302
		Large section of complete individual, showing crust and one polished surface, 15 x 40 cm. The crust surface is smooth, but the usual pittings are absent. Bead-like projections mark the presence of metallic nodules which resisted fusion. Cat. No., 347.	11,365
		Full-sized slab, 20 x 33 cm., polished. The metallic veins first described by Preston, are beautifully exhibited in this specimen. Cat. No., 348.	7,792
		Nearly complete individual. Metallic beads are numerous on the surface, and the scale-like crust seems to be largely metallic. In other respects, like previous specimen. Cat. No., 349.	5,494
		Section showing natural and polished surfaces. Cat. No., 343.	327
86	Fell 1849, Oct. 31, 3 P. M.	**Flows,** Cabarrus Co., N o r t h C a r o l i n a. This fall is often known as Monroe, Cabarrus Co. Monroe is, however, not in Cabarrus Co., and eighteen miles distant. As the fall took place near the present post-office of Flows, this seems to be a more suitable name by which to designate the fall than Monroe.	

No.	Date of Fall or Find.	NAME AND DESCRIPTION.	Weight in grams.
		Stone. Veined gray chondrite, Cga. Fragment from interior. Dark-gray, with white, rounded chondri and numerous metallic grains. Compact. Cat. No., 256.	
		Floyd County, see Indian Valley.	4
87	Fell 1890, May 2, 5:15 P. M.	**Forest City,** Winnebago Co., I o w a. Stone. Spherical chondrite, Cc. 677 complete individuals, ranging in weight from one-tenth of an ounce to one pound. They exhibit almost every variety of shape and degree of surface fusion. From the fully rounded specimens with thick, black crust there is every gradation to those whose rough surface is only slightly blackened, indicating that they separated from other masses only a short distance before reaching the earth. The interior, where seen, is light-gray, with coarse metallic particles. In the group is the stone which fell into a hay-stack without setting it on fire. Cat. Nos., 340-41, 326.	
			10,452.75
		Complete individual, with small, conical pittings resembling raindrop impressions. Cat. No., 340.	4,308.25
		Complete individual, with small, conical pittings resembling raindrop impressions. Cat. No., 340.	1,698
88	Found 1891.	**Forsyth County,** N o r t h C a r o l i n a. Iron. Ataxite, Nedagolla group. Polished slab. Cat. No., 568.	384
89	Fell 1829, May 8, 3:30 P. M.	**Forsyth,** Monroe Co., G e o r g i a. Stone. Veined white chondrite, Cwa. Thin chip, polished. Ground-mass brownish-gray, containing chondri of lighter color, and scattered fine metallic grains. Also smaller fragments. Cat. No., 240.	5
		Fragment with crust. Crust black, dull, and thick. Interior like previous specimen. Cat. No., 241.	34
90	Found 1882.	**Fort Duncan,** Maverick Co., T e x a s. Iron. Hexahedrite H. Thin slab, with crust etched. The etched surface is stippled and shows Neumann lines. Small grains of troilite are visible. Cat. No., 113.	104
91	Fell 1868, Dec. 5, 3 P. M.	**Frankfort,** Franklin Co., A l a b a m a. Stone. Howardite, Ho. Thin, sawed fragment. Light-gray with black and white grains. No metallic particles visible. Cat. No., 296.	0.5
92	Recognized 1889.	**Gilgoin,** Gilgoin Station, N e w S o u t h W a l e s. Stone. Crystalline chondrite, Ck. Polished fragment, with crust. Polished surface, with crust, compact, abundantly sprinkled with metallic grains. Light-colored chondri, 1 mm. in diameter are to be seen. The crust surface is rough and rusty in appearance. Cat. No., 538.	214

No.	Date of Fall or Find.	NAME AND DESCRIPTION.	Weight in grams.
93	Fell 1853, Feb. 10, 1 P. M.	**Girgenti,** Sicily. Stone. Veined white chondrite, Cwa. Polished fragments from interior. Gray, very fine grained, with bright metallic grains. Cat. No., 263.	1.1
94	Found 1884.	**Glorieta Mountain,** Santa Fe Co., New Mexico. Iron. Medium octahedrite, Om. Slab, 11 x 13 cm., with crust, polished and etched. The well-known Widmanstätten figures of this iron are fully displayed. Cat. No., 122. Square section etched, showing both coarse and fine Widmanstätten figures. Cat. No., 123.	1,271 13
95	Found 1883.	**Grand Rapids,** Kent Co., Michigan. Iron. Fine octahedrite, Of. Full-sized section, 13 x 17 cm., polished and etched. Shows very distinct and striking Widmanstätten figures made up of thin plates packed together in bundles. Cat. No., 116. Full-sized, thick section, 16 x 22 cm., polished and showing Widmanstätten figures like previous specimen. Cat. No., 117.	1,160.5 7,881
96	Fell 1861, June 28, 7 P. M.	**Grosnaja** (Mikenskoi), Caucasus, Russia. Stone. Black chondrite, Cs. Fragments from interior. Color, black, spotted with white chondri. Occasional metallic grains. Texture compact. Cat. No., 485.	4
97	Fell 1837, July 24, 11:30 A. M.	**Gross-Divina,** Trentsiner Com., Hungary. Stone. Spherical chondrite, Cc. Fragments of interior and crust. Interior brownish-gray, friable. Crust black. Cat. No., 545.	0.56
98	Fell 1881, Nov. 19, 6:30 A. M.	**Grossliebenthal,** near Odessa, Gov. Kherson, Russia. Stone. Veined white chondrite, Cwa. Fragments, with crust. Crust black, rather smooth. Interior light-gray, fine-grained, rather friable. Shows shining metallic grains and a narrow vein. Cat. No., 549.	0.91
99	Found 1856.	**Hainholz,** Minden, Westphalia. Iron-stone. Mesosiderite, M. Thin chip, showing natural and polished surface. The metallic grains are small, and scattered through a brownish mass of asmanite and bronzite. Cat. No., 165. Fragment from interior. Black, fine-grained. Cat. No., 267.	10.5 1.3
100	Found 1884.	**Hammond,** St. Croix Co., Wisconsin. Iron. Compact iron with octahedral streaks, Do. Thin slab, showing one etched and one polished surface. The component granules have a peculiar, shagreened appearance, due to their grouping in smaller and larger squares and to scattered flakes of schreibersite. Cat. No., 124.	35

No.	Date of Fall or Find.	NAME AND DESCRIPTION.	Weight in grams.
		Hartford, see Marion.	
101	Found 1895.	**Hayden Creek,** Lemhi Co., I d a h o. Iron. Medium octahedrite, Om. Etched section with crust. Shows typical octahedral figures. Cat. No., 489.	51
		Henry County, see Hopper.	
102	Fell 1857, April 1, Night.	**Heredia,** Costa Rica, C e n t r a l A m e r i c a. Stone. Brecciated spherical chondrite, Ccb. Fragment from the interior. Light-gray with metallic grains. Texture rather firm. Cat. No., 442.	4,5
103	Fell 1869, Jan. 1, 12:30 P. M.	**Hessle,** near Upsala, S w e d e n. Stone. Spherical chondrite, Cc. Sawed fragment, with thin, dull black crust. Metallic grains coarse and numerous. Cat. No., 297. Fragment showing crust on all but two surfaces. Cat. No., 298.	18 5
104	Found 1882.	**Hex River,** Cape Colony, S o. A f r i c a. Iron. Hexahedrite, H. Sawed slab, one surface etched. Neumann lines are partially discernible, but more prominent are the parallel systems of troilite plates described by Brezina. These are beautifully shown in this specimen. Cat. No , 115.	364
105	Found 1887.	**Holland's Store,** Chattooga Co., G e o r g i a. Iron. Brecciated hexahedrite, Hb. Thin fragment with crust. Polished surface. Cat. No., 129.	28
106	Fell 1875, Feb. 12, 10:15 P. M.	**Homestead,** Iowa Co., I o w a. Stone. Brecciated gray chondrite, Cgb. Complete individual, nearly covered with crust. Surface indented with broad, shallow pits. Crust thin, dull black. Interior of stone dark-gray. Cat. No., 312. About three-fourths of a complete individual. Crust and interior like previous specimen. The chondritic structure is well exhibited, and metallic grains are numerous. Cat. No., 313. Polished slab, 16 x 20 cm. with crust. The abundance of metallic constituents is well displayed in this specimen as are also the chondri. Cat. No., 314.	3,175 7,626 1,744
107	Prehistoric. Described 1902.	**Hopewell Mounds,** Ross Co., O h i o. Iron. Medium octahedrite, Om. Fragment with etched surface. Cat. No., 480. Fragments, worked into beads. Cat. No., 481.	125.5 11.5

No.	Date of Fall or Find.	NAME AND DESCRIPTION.	Weight in grams.
108	Found 1889.	**Hopper,** Henry Co., V i r g i n i a. This meteorite is usually known by the name of Henry County, but as the place of find is accurately given near the present post-office of Hopper, the above name seems desirable. Both Wülfing and Berwerth inquire whether this meteorite should not be classed with Locust Grove, but the two bear no resemblance in structure and the localities are three hundred miles apart. The Hopper iron somewhat resembles, however, that of Smith's Mountain in structure and composition and the localities are only eighteen miles apart. Iron. Medium octahedrite, Om. Cleavage pieces, (octahedral) much oxidized. Cat. No., 136.	47
109	Found about 1800.	**Imilac,** Atacama, C h i l e. Iron-stone. Pallasite, P. Fragment of iron matrix, most of the stony filling having dropped out. Cat. No., 160. Like previous specimen. Cat. no., 161. Thick slab, polished and etched. The metallic portion exhibits occasional Widmanstätten figures. Its sponge-like pores are filled with chrysolite more or less decomposed. Cat. No., 162.	12.5 28.5 205
110	Found 1888.	**Inca** (Llano del Inca), Atacama, C h i l e. Iron-stone. Mesosiderite, M. Dark brown mass, with natural and polished surfaces. Metallic grains appear only on one edge. Cat. No., 188. Complete individual, intersected by the cracks so characteristic of this meteorite. A few large grains of chrysolite are enclosed in cavities on the surface. Cat. No., 189. Thick slab, polished on two surfaces. No metallic grains are visible. Cat. No., 190.	38 54.5 148
111	Fell 1891, Apr. 7.	**Indarch,** Caucasus, R u s s i a. Stone. Carbonaceous spherical chondrite, Kc. Fragment from interior. Color, brownish black. Appears porous on account of chondri having dropped out. Numerous chondri, some of them shiny black, all spherical and up to 1.5 mm. in diameter, project from the mass. Minute, shining, metallic grains are numerous and show in one place a vein-like arrangement. Cat. No., 506.	15
112	Found 1887.	**Indian Valley,** Indian Valley Township, Floyd Co., V i r g i n i a. Iron. Hexahedrite, H. There seems to be no evidence of brecciation in this iron sufficient to warrant calling it a brecciated hexahedrite. About half of the original mass, one surface polished. This surface is of bright nickel-white color and exhibits numerous parallel rows of rhabdite inclusions. The crust surface shows rather deep pittings. Cat. No., 154.	7,426

No.	Date of Fall or Find.	NAME AND DESCRIPTION.	Weight in grams.
		Full-sized section, 9 x 16 cm., etched. The etched surface exhibits a bright sheen flecked with granular spots arranged in lines. Two small nodules of troilite are visible. See Plate XXXVI. Cat. No., 463.	659
113	Found 1880.	**Ivanpah,** San Bernardino Co., C a l i f o r n i a. Iron. Medium octahedrite, Om. Chiseled fragment which shows no evidence of cleavage. Cat. No., 112.	3
114	Found 1885.	**Jamestown,** Stutsman Co., N o r t h D a k o t a. Iron. Fine octahedrite, Of. Etched section, 2.5 x 11 cm., with crust. Widmanstätten figures are but dimly outlined. Cat. No., 483.	104
115	Found 1885.	**Jamyschewa** (Pawlodar), Semipalatinsk, R u s s i a. Iron-stone. Pallasite, P. Polished fragment, made up principally of chrysolite enclosed in an iron matrix. Also some loose grains of chrysolite. Cat. No., 183.	15
116	Fell 1889, Dec. 1, 2:30 P. M.	**Jelica,** S e r v i a. Stone. Amphoterite, Am. Fragment with crust. Crust black, rather smooth, the surface showing typical pittings. Interior gray, with angular pieces of bronzite projecting from a fine-grained ground-mass. Texture friable. Metallic grains small and scarce. An excellent specimen. Cat. No., 511.	69
117	Found 1883.	**Jenny's Creek,** Wayne Co., W e s t V i r g i n i a. Iron. Coarse octahedrite, Og. Chiseled fragment, showing cleavage octahedrons, separated by bright plates of taenite. Cat. No., 114.	21
118	Fell 1873, June.	**Jhung,** Punjab, I n d i a. Stone. Spherical chondrite, Cc. Fragment from interior. Grayish brown, coarse-grained, chondritic, metallic particles few and small. Cat. No., 305. Thin, polished fragment, showing characters like previous specimen. Cat. No., 306.	4 2.5
119	Found 1884.	**Joe Wright Mountain,** Independence Co., A r-k a n s a s. Iron. Medium octahedrite, Om. Thin slab, etched, showing nodules of troilite and typical Widmanstätten figures. The arrangement of plates about one of the troilite nodules suggests a spherulite. Cat. No., 120.	98.5
120	Found 1866.	**Juncal,** Atacama, C h i l e. Iron. Medium octahedrite, Om. Thin slab with crust, etched. Shows well-marked Widmanstätten figures, the plates of taenite being very distinct. Cat. No., 140.	60

No.	Date of Fall or Find.	NAME AND DESCRIPTION.	Weight in grams.
121	Fell 1821, June 15, 3:30 P. M.	**Juvinas,** Dep. Ardêche, F r a n c e. Stone. Eukrite, Eu. Three fragments from interior. Dark-gray. Structure not well defined. No metallic grains visible. Cat. No., 237.	12
122	Known 1887.	**Kendall County,** T e x a s. Iron. Brecciated hexahedrite, Hb. Thin slab with natural, sawed and etched surfaces. The etched surface exhibits a coarsely granular structure, crossed by a network of delicate, straight lines. Shows numerous nodules of troilite, one 15 mm. in diameter. Cat. No., 138.	118
123	Found 1889.	**Kenton County,** K e n t u c k y. Iron. Medium octahedrite, Om. About one-third of the original mass, showing crust and polished surface, 22 x 42 cm. Contains nodules of troilite. Cat. No., 133.	36,600
		Full-sized slab, 14 x 49 cm., etched. Widmanstätten figures very distinct and regular. Shows marked cleavage and tendency to separate along the cleavage planes. Very perfect octahedrons can be cleaved out from the mass. Cat. No., 134.	7,483
		Full-sized slab, 23 x 44 cm. Both sides polished. Cat. No., 135.	12,231
124	Fell 1869, May 22, 10 P. M.	**Kernouve** (Cléguérec), Morbihan, F r a n c e. Stone. Crystalline chondrite, Ck. Fragment from interior. Cat. No., 299.	0.4
		Thin chip, one surface polished. The polished surface is grayish-brown in color, dotted with minute metallic grains, and large black chondri ½ mm. in diameter. The stone takes an excellent polish. Cat. No., 300.	2
		Thin, polished fragment. Cat. No., 301.	24
125	Fell 1850, June 13.	**Kesen,** Iwate Prefecture, J a p a n. Stone. Brecciated spherical chondrite, Ccb. Mass showing crust and interior. The crust surface differs little from the interior except that the metallic grains of the former have been blackened by fusion, and broad, shallow pits appear on this surface. The interior is dark-gray, compact and plentifully sprinkled with rusty iron grains. A portion of the surface is smoothed and grooved, indicating slipping along these planes. Cat. No., 257.	1,286
		Similar to previous specimen, but showing elongated pits on the crust surface. Cat. No., 258.	1,211
126	Fell 1866, June 9, 5 P. M.	**Knyahinya,** near Nagy-Berezna, H u n g a r y. Stone. Gray chondrite, Cg. One-half of a complete individual, showing crust and polished surface, 13 x 18 cm. The latter exhibits large and small chondri, with few metallic grains. Cat. No., 284.	82

No.	Date of Fall or Find.	NAME AND DESCRIPTION.	Weight in grams.
		Complete individual, covered with thin, black crust. Cat. No., 285.	10
		Complete individual, mostly covered with black, somewhat shining crust. Surface indented with shallow pits. Cat. No., 286.	239
		Flattened mass, showing crust and one polished surface. The crust surface is smooth and covered with small, conical pittings, giving to it the appearance of having a cellular structure. The polished surface well exhibits the aggregation of chondri which make up the mass of the stone. Some of the chondri reach a diameter 3 mm. Cat. No., 287.	3,231
		Complete individual, of irregular, pyramidal form, surface covered with shining black crust. Cat. No., 288.	82.5
127	Found 1749.	**Krasnojarsk** (Medwedewa), Gov. Jeniseisk, S i - b e r i a.	
		The Pallas Iron. Iron-stone. Pallasite, P. Fragment of the iron matrix with a little chrysolite. Cat. No., 157.	9
		Chiseled fragment showing both iron and chrysolite. Cat. No., 158.	12.5
		Several fragments, composed of iron and chrysolite. Cat. No., 159.	76.5
	Found 1891.	Sawed fragment with crust, etched. Shows typical octahedral figures. Cat. No., 505.	169
		Melnikoff regards this a separate fall.	
128	Recognized 1828.	**La Caille,** Dep. Var, F r a n c e.	
		Iron. Medium octahedrite, Om. Etched slab with crust. Displays the typical swollen octahedral figures of this meteorite. Cat. No., 541.	33
129	Found 1860.	**La Grange,** Oldham Co., K e n t u c k y.	
		Iron. Fine octahedrite, Of. Sawed section with crust, and etched surface. The short, narrow, irregular bands of kamacite are bordered by thin plates of taenite. Cat. No., 84.	47
130	Fell 1803, April 26, 1 P. M.	**L'Aigle,** Normandie, Dep. Orne, F r a n c e.	
		Stone. Brecciated intermediate chondrite, Cib. Gray powder. Cat No., 218.	.25
		Fragment with crust. The latter thin, reddish brown, smooth. Interior grayish-brown, compact, porphyritic in appearance. Cat. No., 219.	111
		Fragment with crust and polished surface. The polished surface shows a few fine, metallic grains. Through a dark, amorphous groundmass are mingled grayish-white nodules of various sizes. Cat. No., 220.	60

No.	Date of Fall or Find.	NAME AND DESCRIPTION.	Weight in grams.
131	Fell 1897, June 20, 8:30 P. M.	**Lançon,** Bouches-du-Rhone, France. Stone. Brecciated gray chondrite, Cgb. Fragment with crust. Interior gray, of uniform texture, with shining metallic grains. Individual chondri not clearly discernible. There are numerous slickensided surfaces running in various directions. Crust, dirty black, cracked and rough. Cat. No., 526.	85
132	Found 1857.	**Laurens County,** South Carolina. Iron. Fine octahedrite, Of. Thin slab, etched, showing beautiful Widmanstätten figures. The delicate bands, silver-white in color, and intersecting in equilateral triangles, stand out in sharp contrast to the dull-gray of the groundmass. Cat. No., 76.	13
133	Found 1814.	**Lenarto,** Saros, Hungary. Iron. Medium octahedrite, Om. Square slab, showing crust on one side and one etched surface. No Widmanstätten figures. Cat. No., 35.	47
134	Fell 1845, Jan. 25, 3 P. M.	**Le Pressoir,** Louans, Dep. Indre et Loire, France. Stone. Spherical chondrite, Cc. Fragment with crust. Crust black, smooth. Interior light-gray, somewhat rusted. Outlines of chondri are barely descernible. Cat. No., 439.	17.5
135	Fell 1896, April 13, 7:30 A. M.	**Lesves,** Namur, Belgium. Stone. Gray chondrite, Cg. Fragment with crust. Interior gray with shining metallic grains. Individual chondri are visible. Crust black and smooth. Cat. No., 542.	10.6
136	Found 1880.	**Lexington County,** South Carolina. Iron. Coarse octahedrite, Og. Thin slab with crust, etched. Etching divides the surface into irregular grains, but no regular structure is visible. Cat. No., 111.	23.5
137	Found 1834.	**Lime Creek,** Monroe Co., Alabama. Iron. Hexahedrite, H. Thin, polished slab. The etched surface bears intersecting short, straight lines of rhabdite. Cat. No., 38. Worked mass. Cat. No., 39. Section with crust. Cat. No., 434.	25.5 128 32
138	Fell 1813, Sept. 10, 6 A. M.	**Limerick,** Adare, Ireland. Stone. Veined gray chondrite, Cga. Thin chip, polished. Dark-gray, with thickly distributed rusty iron flakes. Cat. No., 233.	1.5

No.	Date of Fall or Find.	NAME AND DESCRIPTION.	Weight in grams.
139	Known 1853.	**Lion River,** Great Namaqualand, S o u t h A f r i c a. Probably the same as Mukerop. Iron. Fine octahedrite, Of. Etched slab, with crust. Beautiful Widmanstätten figures are displayed, the plates being narrow and very distinct. Cat. No., 62.	27
		Like No. 62, but Widmanstätten figures less distinct. Cat. No., 376.	62.5
140	Fell 1808, Sept. 3, 3:30 P. M.	**Lissa,** Bunzlau, B o h e m i a. Stone. Veined white chondrite, Cwa. Fragment from interior. Light-gray, with few metallic grains. White chondri of various sizes are visible here and there. Cat. No., 510.	32.5
141	Fell 1820, July 12, 5:30 P. M.	**Lixna** (Lasdany), Witebsk, R u s s i a. Stone. Veined gray chondrite, Cga. Fragment from interior. Dark-gray with shining metallic grains. Slickensided surface. Cat. No., 550.	1.95
142	Found 1857	**Locust Grove,** Henry Co., G e o r g i a. Iron. Ataxite, Siratik group. Thick slab with crust. Etched. Granular structure. Rhabdite inclusions. Cat. No., 558.	370.5
143	Found 1892.	**Long Island,** Phillips Co., K a n s a s. Stone. Crystalline chondrite, Ck. Nearly complete individual, made up of four pieces which have been placed together along the line of original fracture. Together with these is a large number of smaller pieces, varying in weight from 10,000 grammes to 5 grammes, probably also a part of the same individual at the time it fell to the earth. The surface of the main mass is indented by shallow, elliptical pits, the long axes of which run in parallel directions. The crust is smooth and brown, but in many places coated with a white incrustation of carbonate of lime, derived from the soil in which the stone lay. The interior of the mass shows a very compact, fine-grained texture, with few metallic grains; color, dark-green. The smaller fragments are much rusted by exposure. Cat. No., 350.	528,488
144	Found 1854.	**Madoc,** Hastings Co., Ontario, C a n a d a. Iron. Fine octahedrite, Of. Spheroidal fragment showing natural surface with pittings. Cat. No., 65.	9
		Thin, sawed slab with natural surface. Cat. No., 66.	5
145	Fell 1886, Nov. 10, 3 P. M.	**Maême,** Hislugari, Prov. Satsuma, J a p a n. Stone. Veined white chondrite, Cwa. Fragment with crust. Crust black and smooth. Interior light-gray with rusty spots and metallic grains. No distinct chondri visible. Texture friable. Cat. No., 440.	7.7

No.	Date of Fall or Find.	NAME AND DESCRIPTION.	Weight in grams.
146	Found 1840.	**Magura,** Arva, H u n g a r y. Iron. Coarse octahedrite, Og. Irregular fragment. Cleavage structure prominent. Cat. No., 46. Fragment showing natural and etched surface. No Widmanstätten figures. Cat. No., 47.	137 166.5
147	Found 1852.	**Mainz,** Hesse, G e r m a n y. Stone. Crystalline chondrite, Ck. Polished fragment. Color, dark-brown, with thinly scattered metallic grains. Light-colored chondri are discernible by their outlines. Cat. No., 441.	1.8
148	Fell 1902, Jan. 6, 10 P. M.	**Majalahti,** F i n l a n d. Iron-stone. Pallasite, P. Section 10 x 5 cm., with one polished and etched surface. Shows usual sponge-like structure of pallasites, the large cavities in the nickel-iron being filled with granular chrysolite. Metallic portion nickel-white in color, not oxidized. Widmanstätten figures appear on the nickel-iron. There are two small-sized inclusions of a bronze-yellow magnetic mineral, probably schreibersite. See Plate XXXIII. Cat. No., 562.	137
149	Fell 1863, Dec. 22, 9 A. M.	**Manbhoom,** Bengal, I n d i a. Stone. Amphoterite, Am. Fragments from interior. Light-gray in color. Shows bronzite and a metallic grain. Cat. No., 551.	0.85
150	Fell 1847, Feb. 25, 2:45 P. M.	**Marion,** Linn Co., I o w a. This fall is often known as Hartford, Linn Co. Hartford is, however, a hundred miles distant and not in Linn Co. The name seems to have been applied because one of the first published letters describing the fall was dated from Hartford. As the fall took place only nine miles from Marion and this is the nearest large town as well as the county-seat, it seems to be the best name by which to designate the fall. Stone. Veined white chondrite, Cwa. Mass with crust. The crust, thick and dull black, is intersected by numerous cracks. Interior pearl-gray, abounding in minute iron grains. Delicate lines of fracture, which traverse the specimen, seem to mark slipping zones with slickensided surfaces. Cat. No., 255.	128
151	Fell 1768, Nov. 20, 4 P. M.	**Mauerkirchen,** A u s t r i a. Stone. White chondrite, Cw. Fragment with crust. Two polished surfaces show scattered metallic grains and well-marked chondri. Cat. No., 211.	110

No.	Date of Fall or Find.	NAME AND DESCRIPTION.	Weight in grams.
152	Fell 1870.	**McKinney,** Collin Co., T e x a s. Stone. Black chondrite, Cs. Plano-convex mass, showing crust and polished surface. Crust reddish-brown, about 1 mm. thick. Interior greenish-black, exhibiting scattered metallic grains and outlines of chondri in almost unbroken connection. Cat. No., 355.	72
153	Described 1875.	**Mejillones,** Atacama, C h i l e. Iron. Brecciated hexahedrite, Hb. Thin, polished slab. The nickel-iron is distributed in a fine network and occasional nodules through an amorphous ground-mass. Cat. No., 172.	37
154	Fell 1862, Oct. 7, 12:30 P. M.	**Menow,** Mecklenburg, G e r m a n y. Stone. Crystalline spherical chondrite, Cck. Fragment from interior, made up of coarse, transparent grains with rusty metallic ones, the whole resembling a piece of brown sandstone. Cat. No., 278.	2
155	Fell 1852, Sept. 4, 4:30 P. M.	**Mezö-Madras,** Transylvania, H u n g a r y. Stone. Brecciated gray chondrite, Cgb. Polished fragment from interior. In the dark-brown ground-mass are sharply outlined gray and white chondri, interspersed with bright, minute grains of nickel-iron. Cat. No., 260. Like previous specimen, but showing rough, dull-brown crust, not sharply separated from the interior. Cat. No., 261.	2 4
156	Fell 1889, June 9, 8:30 A. M.	**Mighei,** Gov. Kherson, R u s s i a. Stone. Carbonaceous chondrite, K. Fragment, with crust. Of dark color, somewhat resembling a piece of graphite, and so friable as to soil the fingers. Crust reddish and scoriaceous. Cat. No., 338. Like previous specimen, except that crust is darker. Chondri of lighter color are distributed through the mass. Cat. No., 339.	1.5 4.4
157	Fell 1842, April 26, 3 P. M.	**Milena** (Pusinsko Selo), Croatia, H u n g a r y. Stone. White chondrite, Cw. Fragment from interior. Light-gray, with coarse and fine metallic grains. Shows distinct chondritic structure. Cat. No , 250.	6
158	Found 1857.	**Mincy,** Taney Co., M i s s o u r i. Iron-stone. Mesosiderite, M. Sawed slab, 10 x 13 cm., showing natural and polished surfaces. The metallic and non-metallic minerals are about equally abundant. Cat. No., 167. Like previous specimen except that the silicates are gathered in large nodules in certain portions. Cat. No., 168. Fragment with natural surface. Cat. No., 77.	395 209 4

No.	Date of Fall or Find.	NAME AND DESCRIPTION.	Weight in grams.
159	Known 1804.	**Misteca,** Oaxaca, M e x i c o. Iron. Medium octahedrite, Om. Porous slab, etched. Widmanstätten figures quite distinct. Cat. No., 32.	86
160	Fell 1882, Feb. 3, 4 P. M.	**Mocs,** Kolos, H u n g a r y. Stone. Veined white chondrite, Cwa. Nearly complete individual, cuboidal in form, with solid angles only slightly rounded. Interior grayish-brown in color, with coarse, metallic grains. Cat. No., 322.	179
		Elongated fragment, showing crust on two sides. Narrow, dark veins pass through the mass in several directions. Cat. No., 323.	41
		Six fragments of nearly equal size, showing crust and interior. They have in general a cuboidal form with a prominence of the solid angles. Portions of the interior display a slickensided surface. Cat. No., 324-29.	543
		Complete individual, tetrahedral in form. Entirely covered with thick, black crust, except at one point, where the light-gray interior may be seen. Cat. No., 330.	80.5
		Complete individual, plano-convex in form, the convex surface being evidently the "breast" side. The opposite face shows a thinner crust and rougher surface. Cat. No., 331.	6
		Monroe, see Flows.	
161	Fell 1810, August, Noon.	**Mooresfort,** Tipperary, I r e l a n d. Stone. Veined gray chondrite, Cga. Fragment with crust. Crust black, somewhat shining. Interior, compact, ash-gray. Shows coarse, metallic grains and white chondri. Cat. No., 228.	7
162	Recognized 1893.	**Mooranoppin,** W e s t A u s t r a l i a. Iron. Coarse octahedrite, Og. Polished and etched section with crust. Widmanstätten figures irregular, and blotched with schreiberite. Resembles Oscuro Mts. in figures. Cat. No., 540.	99
163	Fell 1826, May 19.	**Mordvinovka,** Gov. Ekaterinoslaw, R u s s i a. Stone. White chondrite, Cw. Fragment from the interior. Light gray, flecked with rust and shining metallic grains. No distinct chondri are visible. Cat. No., 448.	15
164	Found 1887.	**Morristown,** Hamblen Co., T e n n e s s e e. Iron-stone. Mesosiderite, M. Full-sized, polished section, 8 x 14 cm. The stony and metallic portions are about equal in quantity. The nickel-iron tends to gather in large, rounded nodules. The stony portion is black and unindividualized megascopically. Cat. No., 561.	407

No.	Date of Fall or Find.	NAME AND DESCRIPTION.	Weight in grams.
165	Found 1887.	**Mount Joy,** Mount Joy Township, Adams Co., Pennsylvania. Iron. Brecciated hexahedrite, Hb. This is classed by Berwerth as Ogg, but to the writer it has the characters of a cubic iron. Thick, etched section 10 x 13 cm., with crust. The well-known fragmental structure of this meteorite is plainly exhibited. Along the lines of union of the fragments flakes and grains of schreibersite are to be seen. Cat. No., 432.	733
166	Found 1868.	**Mount Vernon,** Christian Co., Kentucky. Iron-stone. Pallasite, P. Polished slab 15 x 15 cm. Shows network of nickel-iron holding rounded to angular masses of chrysolite. The nickel-iron is unequally distributed, occurring now in a solid mass, now in a network, and now disappearing entirely. It frequently contains inclusions of schreibersite. The chrysolite is of dark-yellow color, transparent to opaque. See Plate XXXIII. Cat. No., 567.	1,000
167	Found 1899.	**Mukerop,** near Tress, in Gibeon region, Ger. S. W. Africa. Iron. Finest octahedrite, Off. Etched slab, showing well-defined, typical Widmanstätten figures. Cat. No., 552. Etched slab, 11 x 7 cm., showing different figures on two sides of a median line. See Plate XXXV. Cat. No., 569.	40.5 429
168	Found 1897.	**Mungindi,** Queensland, Australia. Iron. Fine octahedrite, Of. Full-sized, etched section, 21 x 9 cm. Widmanstätten figures well marked. Numerous troilite inclusions. See Plate XXXVII. Cat. No., 461.	627
169	Found 1847.	**Murfreesboro,** Rutherford Co., Tennessee. Iron. Medium octahedrite, Om. Etched slab showing distinct Widmanstätten figures, the plates of which run principally at right angles. Cat. No., 58.	20.5
170	Found 1899.	**Murphy,** Cherokee Co., North Carolina. Iron. Hexahedrite, H. Polished and etched section. Shows typical sheen of hexahedrites; also a few small troilite inclusions. Cat. No., 503.	125
171	Fell 1879, July 1, Evening.	**Nagaya,** Entre Rios, Argentina, South America. Stone. Carbonaceous chondrite, K. Small fragment, entirely black in color, one surface having a scoriaceous appearance, the remainder the luster of graphite. Cat. No., 320. Several fragments, having much the appearance of bits of black lava. Cat. No., 321.	10 0.5

No.	Date of Fall or Find.	NAME AND DESCRIPTION.	Weight in grams.
172	Found 1890.	**Nagy=Vazsony,** H u n g a r y. Iron. Medium octahedrite, Om. Thin slab showing natural, etched and polished surfaces. Typical Widmanstätten figures. Cat. No., 139.	37
173	Fell 1825, Feb. 10, Noon.	**Nanjemoy,** Charles Co., M a r y l a n d. Stone. Spherical chondrite, Cc. Fragments from interior. Light-gray, fine-grained, somewhat friable. Metallic particles thickly distributed. Cat. No., 238.	0.5
174	Found 1864.	**Nejed,** Wadee Banee Khaled, C e n t r a l A r a b i a. Iron. Medium octahedrite, Om. Etched slab, with crust. Typical octahedral figures. See Plate XXXII. Cat. No., 523.	180
175	Found 1856.	**Nelson County,** K e n t u c k y. Iron. Coarsest octahedrite, Ogg. Large scaling, slightly oxidized. Cat. No., 73. Large polished slab, upon which coarse Widmanstätten figures are here and there dimly outlined. Cat. No., 488.	23 455
176	Found 1872.	**Nenntmannsdorf,** near Pirna, S a x o n y, G e r - m a n y. Iron. Hexahedrite, H. Cleavage pieces. Cat. No. 447.	3.7
177	Found 1897.	**Ness County,** K a n s a s. Stone. Crystalline chondrite, Ck. Nearly complete individual. Cat. No., 490.	85
178	Fell 1860, May 1, 12:45 P. M.	**New Concord,** Muskingum Co., O h i o. Stone. Veined intermediate chondrite, Cia. Nearly complete individual of flattened, tetrahedral form, angles little rounded. A smooth, somewhat shining, black crust covers the slightly pitted surface. Interior dark-gray, compact, and fine-grained. Metallic grains numerous. Cat. No., 273. Section from flattened individual, showing crust and two polished surfaces. The crust is thin, dull-black to reddish. A vein of metallic matter runs through the mass, and stands out in relief from the crust. The interior of the stone is dark-brown and gray. Metallic grains are large and abundant. Cat. No., 274.	347 753
179	Found 1895.	**Oakley,** Logan Co., K a n s a s. Stone. Crystalline chondrite, Ck. Polished section, 5 x 14 cm., with crust. Firm texture, black color, abundant sprinkling of metallic grains, resembling Pipecreek in this feature and approaching a mesosiderite. Rounded spots ½ mm. in diameter indicate the presence of chondri. Crust surface rusted. Cat. No., 501. Gift of Prof. H. A. Ward.	263

No.	Date of Fall or Find.	NAME AND DESCRIPTION.	Weight in grams.
180	Fell 1887, Aug. 30, 3 P. M.	**Ochansk** (Tabory), Gov. Perm, R u s s i a. Stone. Brecciated spherical chondrite, Ccb. Fragment with crust. The latter about 1 mm. thick, dull-black and blebby. Interior of stone light bluish-gray. Largely made up of distinct chondri. Small metallic grains are numerous. Cat. No., 335.	23.5
181	Known 1856.	**Orange River,** Garib, S o u t h A f r i c a. Iron. Medium octahedrite, Om. Sawed section with natural surface, smooth and deeply pitted. Cat. No., 71. Etched slab, showing typical Widmanstätten figures and nodule of troilite. Cat. No., 72.	114 95.5
182	Fell 1864, May 14, 8 P. M.	**Orgueil,** Dep. Tarn et Garonne, F r a n c e. Stone. Carbonaceous chondrite, K. Coarse, black powder, soft and friable. Cat. No., 282. Fragments similar to No. 282, some of them showing crust. Cat. No., 509.	1 20
183	Fell 1868, July 11.	**Ornans,** Doubs, F r a n c e. Stone. Spherical chondrite, ornansite, Cco. Fragment, sawed from interior. Resembles a lump of hardened, sandy mud. Cat. No., 294.	19.5
184	Found 1895.	**Oscuro Mountains,** Socorro Co., N e w M e x i c o. Iron. Medium octahedrite, Om. Etched section with crust. Resembles Mooranoppin. Widmanstätten figures irregular. Surface spotted with inclusions. Crust surface jagged and torn. See Plate XXXII. Cat. No., 457.	113
185	Fell 1855, May 11, 3:30 P. M.	**Ösel** (Kaande), Island of Ösel, Livonia, R u s s i a. Stone. White chondrite, Cw. Fragments with crust. Interior light-gray with rusty and bright metallic grains. Friable. Crust .5 mm. thick, dull-black, papillated. Cat. No., 264.	2
186	Fell 1857, Feb. 28, Noon.	**Parnallee,** Madras, I n d i a. Stone. Veined gray chondrite, Cga. Fragment with crust. The latter is thin, brownish-black and differs little from the rest of the stone. The interior is coarse-grained, with few metallic grains. Cat. No., 270. Fragment with crust, one surface polished. The polished surface is mottled with gray and white chondri of various sizes. Large grains of troilite are visible and minute nickel-iron grains. This meteorite should be classed as an intermediate chondrite. Cat. No., 437.	3.5 167
187	Fell 1863, Aug. 8, 12:30 P. M.	**Pillistfer,** Livonia, R u s s i a. Stone. Crystalline chondrite, Ck. Fragment from interior. Dark-gray, compact. Made up of dark, transparent grains with a large number of minute specks of troilite. Cat. No., 279.	1

No.	Date of Fall or Find.	NAME AND DESCRIPTION.	Weight in grams.
188	Found 1887.	**Pipecreek,** Brandera Co., T e x a s. Stone. Crystalline chondrite, Ck. Irregular fragment, with one polished surface. A dark, heavy stone, with a large proportion of metallic grains. Cat. No., 337.	84
189	Found 1850.	**Pittsburg** (Miller's Run), Allegheny Co., P e n n-s y l v a n i a. Iron. Hexahedrite, H. Etched fragment with crust. No Widmanstätten figures are visible. Cat. No., 433.	4.7
190	Fell 1723, June 22.	**Ploschkowitz,** Bunzlauer District, B o h e m i a. Stone. Brecciated spherical chondrite, Ccb. Oxidized fragments bearing thin, black crust. Cat. No., 493.	2.2
191	Fell 1819, Oct. 13, 8 A. M.	**Politz,** near Gera, Reuss, G e r m a n y. Stone. Veined white chondrite, Cwa. Fragment from interior. Dark-gray, with metallic grains. Cat. No., 236.	0.5
192	Recognized 1893.	**Prairie Dog Creek,** Decatur Co., K a n s a s. Stone. Spherical crystalline chondrite, Cck. Fragment from interior. Texture compact. Color rusty-brown with shining, minute metallic grains. A distinct chondrus, 1 mm. in diameter, is almost separated from the mass. Cat. No., 563.	25
193	Fell 1868, Jan. 30, 7 P. M.	**Pultusk,** Sielce, Nowy, Gostkówo, etc., P o l a n d. Stone. Veined gray chondrite, Cga. Part of a large individual, showing crust and interior. The former, dull-black, papillated; the latter, gray with rusty iron grains. All fine-grained. Cat. No., 289. 49 complete individuals, varying in size from a pea to a walnut. All covered more or less with crust, in some cases showing complete fusion of the surface, in others only a smoking of the same. Cat. No., 290. Seven complete individuals of larger size than previous specimens. Covered with crust. Cat. No., 291–2.	350 740 506
194	Found 1885.	**Puquios,** C h i l e. Iron. Medium octahedrite, Om. Full-sized slab, etched. Irregular Widmanstätten figures are dimly brought out by the etching, also flakes of schreibersite. Cat. No., 377.	154
195	Found 1839.	**Putnam County,** G e o r g i a. Iron. Fine octahedrite, Of. Cleavage pieces showing octahedral form, separated by thin plates of taenite. Cat. No., 44.	4.5

No.	Date of Fall or Find.	NAME AND DESCRIPTION.	Weight in grams.
196	Found 1882.	**Rancho de la Pila,** Durango, M e x i c o. Iron. Medium octahedrite, Om. Etched slab, 6.5 x 11 cm. with crust. Exhibits a coarsely granular structure, but no octahedral figures. There are numerous partings, some of which are quite open. Cat. No., 521.	180
197	Found 1810.	**Rasgata,** Zipaquira, U. S. of C o l o m b i a. Iron. Nickel-poor ataxite, Nedagolla group, Dn. Thin, polished fragment. Cat. No., 435.	2
198	Found 1808.	**Red River,** T e x a s. (Gibbs meteorite.) Iron. Medium octahedrite, Om. Chiseled fragment, one end etched. Symmetrical Widmanstätten figures are shown. Cat. No., 34.	55
199	Found 1895.	**Reed City,** Osceola Co., M i c h i g a n. Iron. Coarse octahedrite, Og. Etched section with crust. Shows coarse Widmanstätten figures and elongated and branching inclusions of schreibersite. Cat. No., 560.	137
200	Fell 1828, June 4, 8:30 A. M.	**Richmond,** Henrico Co., V i r g i n i a. Stone. Crystalline spherical chondrite, Cck. Fragment from interior. Composed chiefly of dark, angular, vitreous and coarse metallic grains. Cat. No., 239.	2
201	Found 1892.	**Roebourne,** Queensland, A u s t r a l i a. Iron. Medium octahedrite, Om. Full-sized, etched section, 22 x 10 cm. Typical octahedral figures are dimly outlined. Cat. No., 460.	1,480
202	Found 1896.	**Sacramento Mountains,** Eddy Co., N e w M e x i c o. Iron. Medium octahedrite, Om. Full-sized, etched section, 12 x 40 cm. Shows typical, well-defined Widmanstätten figures, and two large nodules of troilite, one of the latter perforated. See Plate XXXVI. Cat. No., 465.	2,330
203	Found 1888.	**Saint Genevieve County,** M i s s o u r i. Iron. Fine octahedrite, Of. Thick slab 12.5 x 9.5 cm., etched. Typical octahedral figures of great clearness and regularity are exhibited. There are two small inclusions of troilite. See Plate XXXVI. Cat. No., 512. Gift of Prof. H. A. Ward.	790
204	Fell 1898, Nov. 15, 9:30 P. M.	**Saline,** Saline Township, Sheridan Co., K a n s a s. Stone. Crystalline spherical chondrite, Cck. Larger part of complete individual. Crust black, with metallic points and one large globule of metal. Interior compact, greenish-black in color, and shows abundant metallic grains. See Plate XXX. Cat. No., 527.	19,500

No.	Date of Fall or Find.	NAME AND DESCRIPTION.	Weight in grams.
		Nearly full-sized section, 13 x 17 cm., one surface polished. Metallic grains show tendency to arrangement in veins. Outlines of abundant small chondri are discernible. See Plate XXXI. Cat. No., 565.	675
205	Found 1897.	**San Angelo,** Tom Green Co., T e x a s. Iron. Medium octahedrite, Om. Full-sized, etched section, 12 x 27 cm. Shows typical octahedral figures, with circular and elongated inclusions of troilite, the latter often distributed in a vein-like manner. Cat. No., 478.	1,501
206	Recognized 1887.	**San Emigdio Range,** San Bernardino Co., C a l i f o r n i a. Stone. Spherical chondrite, Cc. Thirteen fragments from interior, one bearing crust. Rusty-brown, with metallic grains. Crust black. Cat. Nos., 446, 451.	12.5
207	Known 1883.	**São Julião de Moreira,** Minho, P o r t u g a l. Iron. Brecciated hexahedrite, Hb. Etched fragment, with crust. Cat. No., 536. Full-sized, polished section, 13 x 27 cm., with crust. Shows coarse, irregular inclusions of schreibersite. See Plate XXXIV. Cat. No., 556.	30 1,782
208	Fell 1868, Sept. 8, 2:30 A. M.	**Sauguis,** Dep. Basses-Pyrénées, F r a n c e. Stone. Veined white chondrite, Cwa. Fragment with crust and polished surface. Crust black and shining, about 1 mm. in thickness. Interior brownish-gray, with scattered metallic particles. Cat. No., 295.	4
209	Fell 1846, Dec. 25, 2:45 P. M.	**Schönenberg,** Swabia, B a v a r i a.| Stone. Veined white chondrite, Cwa. Fragment with crust. The latter is thick, somewhat shining and scoriaceous. The interior is dark-gray, shows metallic grains and light and dark chondri and is traversed by narrow, branching veins of nickel-iron. Cat. No., 254.	8.2
210	Found 1867,	**Scottsville,** Allen Co., K e n t u c k y. Iron. Hexahedrite, H. Full-sized slab, 12 x 17 cm., etched. Contains a circular nodule of troilite. The etched surface has the appearance of a network of delicate, straight lines overlaying a granular base. Cat. No., 91.	364
211	Fell 1871, May 21, 8:15 A. M.	**Searsmont,** Waldo Co., M a i n e. Stone. Spherical chondrite, Cc. Fragment from interior. Light-gray. Cat. No., 302. Various fragments from interior. Light-gray, with metallic grains of silvery lustre. Chondritic structure. Cat. No., 303.	0.25 3

No.	Date of Fall or Find.	NAME AND DESCRIPTION.	Weight in grams.
212	Found 1847.	**Seeläsgen,** Brandenburg, Prussia. Iron. Coarsest octahedrite, Ogg. Chiseled fragment. No cleavage structure visible. Cat. No., 57. Etched slab, containing large nodule of troilite. The iron is seen to be made up of large irregular plates. Cat. No., 375.	41.5 12.5
213	Fell 1773, Nov. 17, Midnight.	**Sena** (Sigena), Aragon, Spain. Stone. Brecciated gray chondrite, Cgb. Fragment from the interior. Compact, mottled white, gray, and brown from presence of chondri and rusty metallic grains. Cat. No., 445.	5.2
214	Found 1850.	**Seneca Falls,** Seneca Co., New York. Iron. Medium octahedrite, Om. Sawed section showing natural surface and fracture. Octahedral cleavage very distinct. One surface partially etched, bears an initial of the name of the first owner, Mr. L. C. Partridge. Cat. No., 60. Loaned by Gen. G. Murray Guion.	300
215	Fell 1865, Aug. 25, 11 A. M.	**Senhadja,** Aumale, Algiers, Africa. Stone. Veined white chondrite, Cwa. Slice from interior. Ash-gray, few metallic grains. Chondritic structure. Cat. No., 283.	1.5
216	Fell 1794, June 16, 7 P. M.	**Siena,** Tuscany, Italy. Stone. Howarditic chondrite, Cho. Fragments from interior. Dark-gray, with no metallic grains visible. Cat. No., 547.	0.25
217	Found 1887.	**Silver Crown,** Laramie Co., Wyoming. Iron. Coarse octahedrite, Og. Etched slab with crust. Structure coarsely crystalline, with a few rectilinear figures. Lines of taenite very distinct. Cat. No., 130.	71
218	Found 1863.	**Smith's Mountain,** Rockingham Co., North Carolina. Iron. Fine octahedrite, Of. Section with crust, etched. Well marked Widmanstätten figures. Some of the bands are of oval shape. Cat. No., 85. Large etched section. Cat. No., 452.	17 231
219	Found 1840.	**Smithville** (Caryfort), Dekalb Co., Tennessee. Iron. Coarse octahedrite, Og. Thin slab, one surface etched, but showing no Widmanstätten figures. Troilite nodules and flakes of schreibersite appear on the etched portion. Cat. No., 50.	55
220	Fell 1877, Oct. 13, 2 P. M.	**Sokobanja,** near Alexinac, Servia. Stone. Spherical chondrite, Cc. Irregular fragment of light-gray color, showing chondri, some of which are 2 mm. in diameter, also grains of troilite and nickel-iron. Friable. Cat. No., 319.	33

No.	Date of Fall or Find.	NAME AND DESCRIPTION.	Weight in grams.
221	Fell 1876, June 28, 11:30 A. M.	**Ställdalen,** S w e d e n. Stone. Brecciated gray chondrite, Cgb. Fragment with crust. The latter black and shining. Interior of the stone dark-gray. Cat. No., 315. Irregular mass, with crust and polished surface. Interior brownish-black. Compact, with numerous metallic grains. There seems to be reason to doubt whether this specimen is really Ställdalen. Cat. No., 316.	3 50
222	Fell 1808, May 22, 6 A. M.	**Stannern,** Moravia, A u s t r i a. Stone. Eukrite, Eu. Fragment from interior. Light-gray. Structure coarse-granular, not chondritic. Cat. No., 225. Fragment with crust; the latter glossy-black, veined. Interior greenish-black, brecciated. Shows one large grain of troilite. Cat. No., 226. Fragment from interior, similar to No. 225. Cat. No., 227.	7.5 23.5 1.5
223	Found 1858.	**Staunton,** Augusta Co., V i r g i n i a. Iron. Medium octahedrite, Om. Full-sized slab, 12 x 23 cm., polished and etched. Shows typical Widmanstätten figures and large nodule of troilite. The latter nearly encloses a portion of nickel iron. Cat. No., 78. Slab with crust, etched. The crystalline plates have an ovoid form, and intersect very irregularly. Cat. No., 79. Slab with crust, polished and etched on two surfaces. Beautiful, broad and distinct Widmanstätten figures. Cat. No., 80.	1,595 665 100.5
224	Found 1724.	**Steinbach** (including Rittersgrün), Saxony, G e r- m a n y. Iron-stone. Siderophyr, S. Thin slab, polished. The stony portion exceeds the metallic. Cat. No., 164.	33.5
	Found 1861.	Breitenbach. Thin, polished slab. Resembles the Steinbach specimen very closely. Cat. No., 169.	1
225	Fell 1753, June 3, 8 P. M.	**Tabor** (Krawin), B o h e m i a. Stone. Brecciated spherical chondrite, Ccb. Fragment from interior. Light-gray with rusty iron spots. Cat. No., 209.	.06
226	Found 1853.	**Tazewell,** Claiborne Co., T e n n e s s e e. Iron. Finest octahedrite, Off. Slab, showing natural and etched surface. Cat. No., 64.	17
227	Found 1886.	**Thunda,** Windorah, Queensland, A u s t r a l i a. Iron. Medium octahedrite, Om. Sawed slab, one surface etched. Widmanstätten figures distinct and regular. Cat. No., 128.	154

No.	Date of Fall or Find.	NAME AND DESCRIPTION.	Weight in grams.
228	Found 1784.	**Toluca,** M e x i c o. Iron. Medium octahedrite, Om. Complete individual. Form spheroidal. Cat. No., 7.	464.5
		Complete individual. Irregular form. Octahedral cleavage well exhibited. Cat. No., 8.	99.5
		Complete individual. Spheroidal form. Cat. No., 9.	263.5
		Complete individual. Shows use as a hammer. Cat. No., 10.	251
		Complete individual. Spheroidal form. Cat. No., 11.	227
		Spheroidal individual, with one etched face showing typical, coarse Widmanstätten figures and nodules of troilite. Cat. No., 12.	816
		Complete individual, showing distinct octahedral cleavage. Cat. No., 13.	112.5
		Complete individual. Cat. No., 14.	225.5
		Crescent-shaped mass, with surface 20 x 40 cm., etched. Shows coarse Widmanstätten figures and nodules of troilite of various shapes and sizes. Cat. No., 15.	16,665
		Similar to above specimen, but smaller. Surface 13 x 20 cm. Cat. No., 16.	6,166
		Broken fragment, showing well developed cleavage planes. Cat. No., 17.	1,997
		Complete individual. Surface pitted and covered with crust. Cat. No., 18.	3,000
		Scalings from previous specimen. Cat. no., 19.	60
		Complete individual. Cat. No., 20.	1,880
		Complete individual, spheroidal. Surface very smooth. Cat. No , 21.	1,107
		Complete individual, hemispheroidal. Cleavage planes well marked. Cat. No., 22.	28,038
		Complete individual, hemispheroidal. Shallow pits appear on the surface. Cat. No., 23.	46,040
		Complete individual, spheroidal. Surface smooth and pitted. Cat. No., 24.	18,025
		Etched section, 18 x 22 cm. The Widmanstätten figures are very distinct and regular. Nodules of troilite of various shapes are included. Cat. No., 25.	1,900
		Like previous specimen, but Widmanstätten figures less distinct. Surface 21 x 38 cm. Cat. No., 26.	2,423
		Complete individual, crescentic in form. Shows strong tendency to scaling and decomposition. Drops of lawrencite appear on the surface. Cat. No., 27.	19,954
		Section of flattened individual, with etched surface, 17 x 17 cm. The latter shows coarse, well-marked Widmanstätten figures and several irregular nodules of troilite. Natural surface deeply pitted. Cat. No., 370.	4,535
		Full-sized slab, 10 x 21 cm., etched. Shows the usual Widmanstätten figures and coarse, vein-like masses of troilite. Cat. No., 371.	823
		Complete individual, showing pittings and natural surface. Form, pyramidal. Cat. No., 372.	2,506

No.	Date of Fall or Find.	NAME AND DESCRIPTION.	Weight in grams.
	Found 1897.	Los Reyes. Complete individual. Small surface etched. Cat. No., 454.	19,500
229	Found 1859-1886.	**Tombigbee River,** Choctaw and Sumter Counties, Alabama. Iron. Ataxite, D. While Berwerth classifies this iron as an ataxite the structure is in many ways cubic. The analysis does not quite accord with those of cubic irons however, and a new analysis is desirable. Etched section 10 x 14 cm., with crust. Shows numerous Neumann lines intersecting at right angles. There are numerous irregular inclusions of schreibersite of nickel-white color. Cat. No., 504.	1,690
230	Found 1863-4	**Tomhannock Creek,** Rensselaer Co., New York. Stone. Brecciated gray chondrite, Cgb. Fragment from interior, polished. Made up chiefly of metallic grains, and a dark-brown, chrysolite-like mineral. Cat. No., 280. Slice, showing crust. Interior portion like previous specimen. Cat. No., 281.	1 7.5
231	Found 1886.	**Tonganoxie,** Leavenworth Co., Kansas. Iron. Medium octahedrite, Om. Full-sized, etched section 8 x 12 cm., with crust. Shows typical octahedral figures with small troilite inclusions bordered by kamacite. Cat. No., 477.	264
232	Fell 1863, Dec. 7, 11 A. M.	**Tourinnes-la-Grosse,** Belgium. Stone. White chondrite, Cw. Fragment from interior. Light-gray, with minute metallic grains. Outlines of chondri not discernible. Friable. Cat. No., 486.	3
233	Found 1858.	**Trenton,** Washington Co., Wisconsin. Iron. Medium octahedrite, Om. Thin slab, etched, showing typical Widmanstätten figures, the plates of which intersect at angles of 35°. Cat. No., 81.	137
234	Fell 1856, Nov. 12, 4 P. M.	**Trenzano,** near Brescia, Italy. Stone. Veined spherical chondrite, Cca. Cubical fragment, with crust on two surfaces. The latter shining, black, only slightly pitted, 3 mm. thick. Interior very compact, coarse-grained, the metallic portion forming a network which encloses dark, spherical chondri, some of a diameter of 2 mm. Cat. No., 268. Smaller fragment, like previous specimen. Cat. No., 325.	57 2
235	Recognized 1851.	**Tucson,** Pima Co., Arizona. Iron. Nickel-rich ataxite, Muchachos group. Etched fragment showing typical stippled appearance of this iron. Cat. No., 59.	12

No.	Date of Fall or Find.	NAME AND DESCRIPTION.	Weight in grams.
236	Fell 1884, May 20, 8:30 P. M.	**Tysnes,** Tysnes Island, N o r w a y. Stone. Brecciated intermediate chondrite, Cib. Fragment from interior. Texture compact and firm. Color grayish-brown, mottled with white and gray. There are also angular inclusions of a lighter color. This stone would seem to be an intermediate chondrite, rather than a gray chondrite, as it is usually designated. Cat. No., 543.	28
237	Found 1853.	**Union County,** G e o r g i a. Iron. Coarsest octahedrite, Ogg. Cleavage fragments with surface considerably oxidized. Cat. No., 63.	1.5
238	Recognized 1861.	**Vaca Muerta** (Sierra de Chaco), C h i l e. Iron-stone. Mesosiderite, M. Fragment with crust. Structure fine-granular, with metallic and non metallic minerals about equally distributed Cat. No., 170. Similar to No. 170, except that the surface appears glazed and shines in iridescent colors. Cat. No., 171.	17.5 14.5
239	Fell 1880, May 1–15.	**Veramin,** Teheran, P e r s i a. Iron-stone. Mesosiderite, M. Fragment from interior. Cat. No., 495.	8
240	Fell 1831, May 13.	**Vouillé,** near Poitiers, Dep. Vienne, F r a n c e. Stone. Veined intermediate chondrite, Cia. Fragment with crust. Interior gray, compact, flecked with rusty iron grains. Several delicate black veins traverse the specimen. Cat. No., 242. Thin chip, polished. Well-marked chondri make up the larger part of the mass. Fine metallic grains are numerous. Cat. No., 243.	53 3
241	Found 1874.	**Waconda,** Mitchell Co., K a n s a s. Stone. Brecciated spherical chondrite, Ccb. Mass from interior. For the most part light-gray in color, the remaining portion harder and darker. Large chondri are visible in the latter portion. Cat. No., 309. Fragment with crust. The latter thin, dull-black, blebby. A dark vein passes through a portion of the specimen. Cat. No., 310. Fragment from interior. Much weathered. Cat. No., 311.	2,835 125 5.5
242	Fell 1877, Jan. 3, 7:15 A. M.	**Warrenton,** Warren Co., M i s s o u r i. Stone. Spherical chondrite, ornansite, Cco. Fragment from interior. Resembles a piece of hardened, sandy mud or blue clay, with a few metallic grains visible. Cat. No., 317.	5

No.	Date of Fall or Find.	NAME AND DESCRIPTION.	Weight in grams.
243	Found 1888.	**Welland,** Ontario, C a n a d a. Iron. Medium octahedrite, Om. Segment, 11.5 x 7.5 cm., showing etched and natural surfaces. Widmanstätten figures distinct and regular. Scattered grains of troilite are present. A marked tendency to octahedral cleavage is apparent. Cat. No., 132.	715.5
244	Fell 1807, Dec. 14, 6:30 A. M.	**Weston,** Fairfield Co., C o n n e c t i c u t. Stone. Brecciated spherical chondrite, Ccb. Fragment with crust. The latter thin, dull-black. Interior divided into yellowish and bluish-gray portions distinctly separated in outline and color. The chondri, of which the mass is largely made up, give it the appearance of a fine conglomerate. Cat. No., 224. Small fragment. Cat. No., 223.	9 3.5
245	Known 1836.	**Wichita County,** T e x a s. Iron. Coarse octahedrite, Og. Etched section, 18 x 27 cm. Shows coarse Widmanstätten figures, nodules of troilite and scattered flakes of schreibersite. Cat. No., 41.	1,396
246	Fell 1795, Dec. 13, 3:30 P. M.	**Wold Cottage,** Yorkshire, E n g l a n d. Stone. Veined white chondrite, Cwa. Three polished chips, showing chondri and metallic grains, both coarse and fine. Cat. No., 215.	2
247	Recognized 1858.	**Wooster,** Wayne Co., O h i o. Iron. Medium octahedrite, Om. Etched fragment, showing typical octahedral figures. Cat. No., 494.	2.7
248	Fell 1852, Jan. 23, 4:30 P. M.	**Yatoor,** near Nellore, Madras, I n d i a. Stone. Spherical chondrite, Cc. Fragments from interior. Gray, with dark chondri and rusty iron grains. Cat. No., 259.	1
249	Found 1884.	**Youndegin,** W e s t e r n A u s t r a l i a. Iron. Coarse octahedrite, Og. Full-sized, elongated slab, 10 x 27 cm., showing pittings, crust, polished and etched surface. The Widmanstätten figures are coarse, many of the plates being 1.5-2 cm. in thickness. They are also crossed by a series of finer plates nearly at right angles. Troilite and schreibersite are present. See Plate XXXVII. Cat. No , 118.	1,087
250	Recognized 1792.	**Zacatecas,** M e x i c o. Iron. Brecciated octahedrite, Obz. Thin fragment, etched. No Widmanstätten figures. Cat. No., 28.	5.7
251	Fell 1897, Aug. 1, 10:30 A. M.	**Zavid,** B o s n i a. Stone. Brecciated gray chondrite, Cgb. Fragment from interior of uniform gray color, with minute, shining metallic grains. Texture rather firm. One surface is slickensided and evidences of brecciation are seen in other portions. Cat. No., 548.	23.5
		Total number falls and finds - 251 Total weight of collection - 2,289,786 grams	

TERRESTRIAL NICKEL-IRONS.

The following Terrestrial Nickel-Irons are catalogued with the collection because of their relation in composition to iron meteorites.

No.	Date of Fall or Find.	NAME AND DESCRIPTION.	Weight in grams.
	Described 1885.	**Gorge River,** Awarua Bay, New Zealand. Sand, containing awaruite. Cat. No., 361.	50
	Described 1892.	**Josephine and Jackson Counties,** Oregon. Josephinite pebble. Cat. No., 367.	1
	Found 1870.	**Ovifak,** Disko Island, Greenland. Crescentic mass, one surface showing polished, homogeneous metal. Cat No., 357. Hemispherical mass, one surface polished. Cat. No., 358. Like no 358. Cat. No., 359.	11,000 1,038 861
	Known 1873.	**Santa Catharina,** Rio San Francisco do Sul, Brazil. The origin of this iron is still in doubt, but to the writer it seems more anomalous as a meteoric than as a terrestrial iron. It seems desirable to class as meteorites only those known to be such, and hence, until confirmation of the meteoric character of this iron can be obtained, it may well be classed as terrestrial. Spheroidal mass, having the well-known limonite-yellow color of the Santa Catharina iron. More or less honey-combed by decay. Cat. No., 97. Similar to previous specimen except that one surface is polished, showing a compact metalloid interior. Cat. No., 99. Similar to No. 97. Cat. No., 100. Similar to No. 97. Cat. No., 103. Mass only slightly altered, of iron-black color and metallic lustre. Cat. No., 98. A number of fragments of various sizes, apparently altered to limonite. Cat. No., 101. Similar to No. 97. Cat. No., 102. Similar to No. 97. Cat. No., 104. Similar to No. 97. Cat. No., 105. Similar to No. 97. Cat. No., 106. Similar to No. 97. Cat. No., 107. Similar to No. 97. Cat. No., 108.	217 921 2,579 4,252 261 1,814 766 3,344 10,884 11,576 3,174 1,577

CASTS OF METEORITES.

About 70 casts or models of meteorites, illustrating the size, form and superficial appearance of original masses, belong to the collection.
The following is a list :

MICRO-SECTIONS.

Thin sections available for microscopic study are possessed of the following falls:

Aussun, Bremervorde, Crab Orchard, Farmington, Forest City, Homestead (3), Kernouve, Kesen, Knyahinya, L'Aigle, Long Island (5), Mocs, Ness County, New Concord, Oakley, Parnallee, Pultusk, Sauguis, Simbirsk and Stannern.

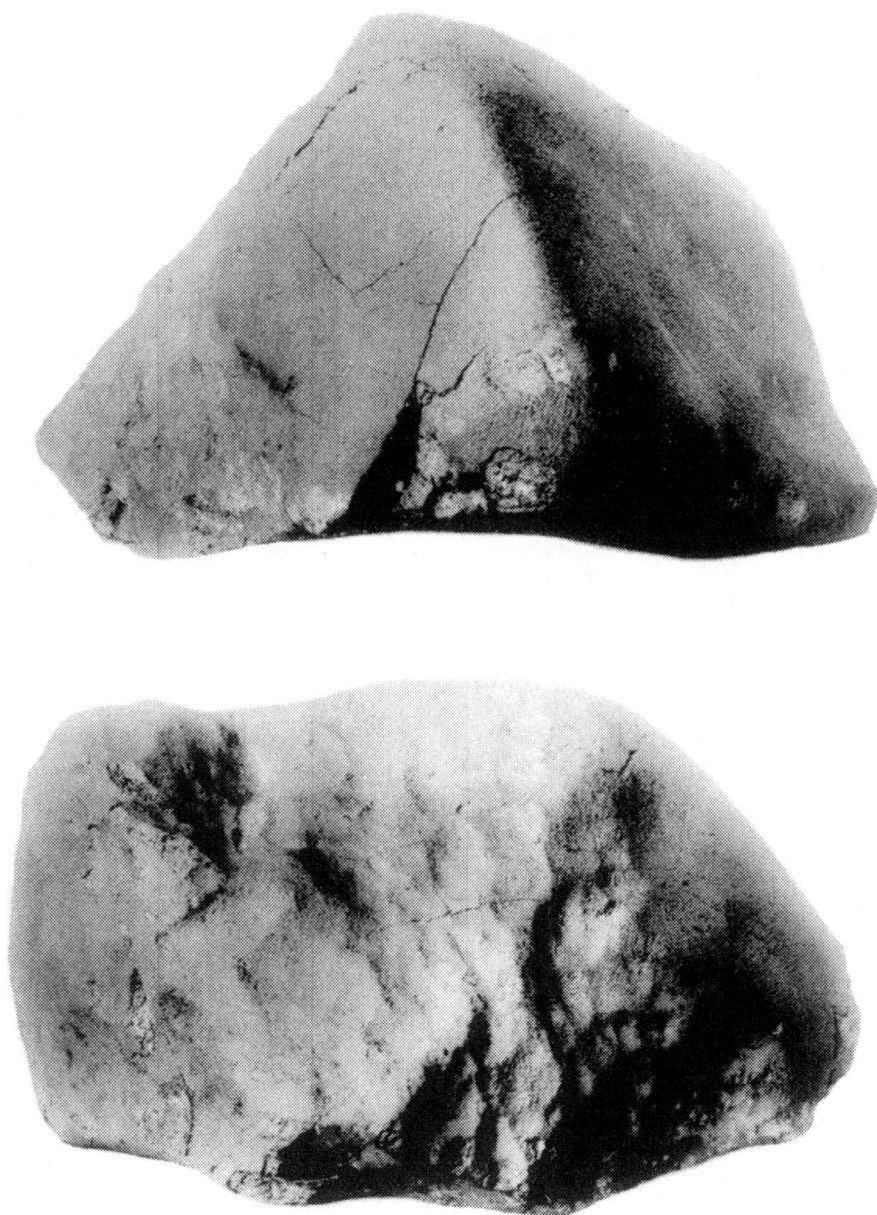

Saline.   Two views of same individual x ⅔.

Saline x ⅔.

Baratta x ⅔.

Bjurböle X ¾.

Bath Furnace X ¾.

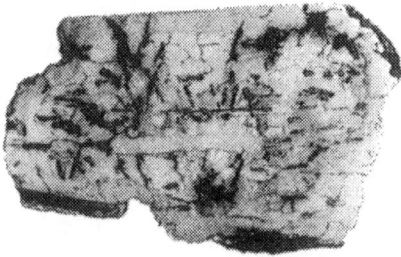

Oscuro Mountains X ⅝

Arlington X ⅝.

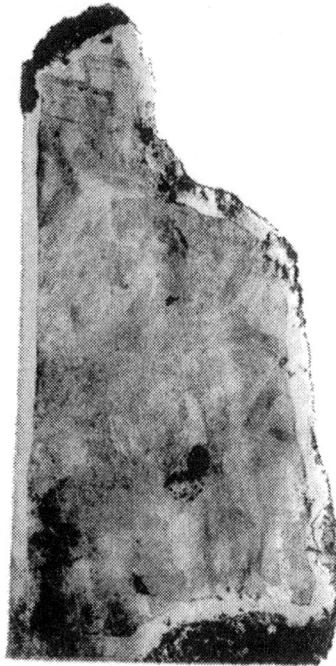

Nejed X ⅝.

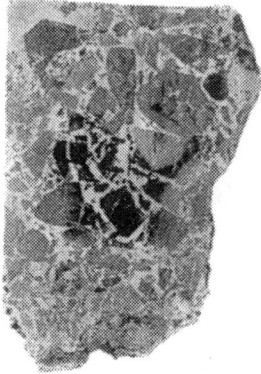

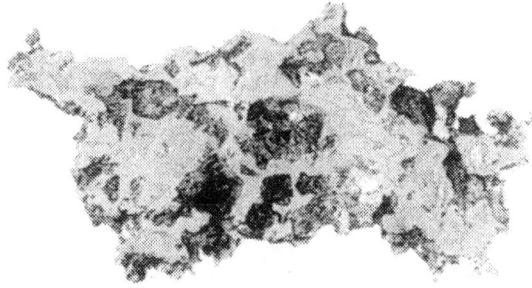

Majalahti x ¾.

Eagle Station x ¾.

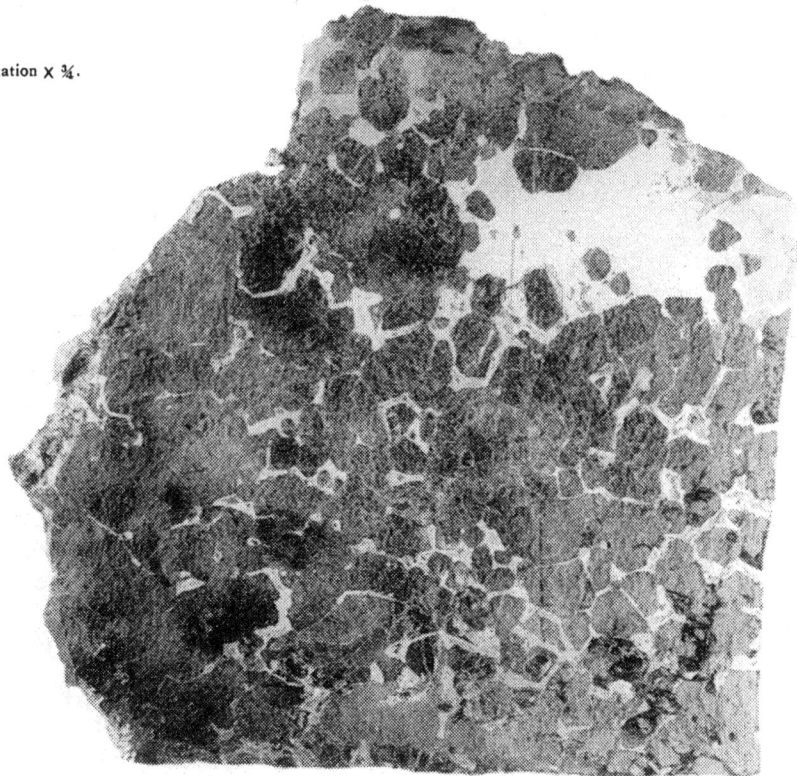

Mount Vernon x ¾.

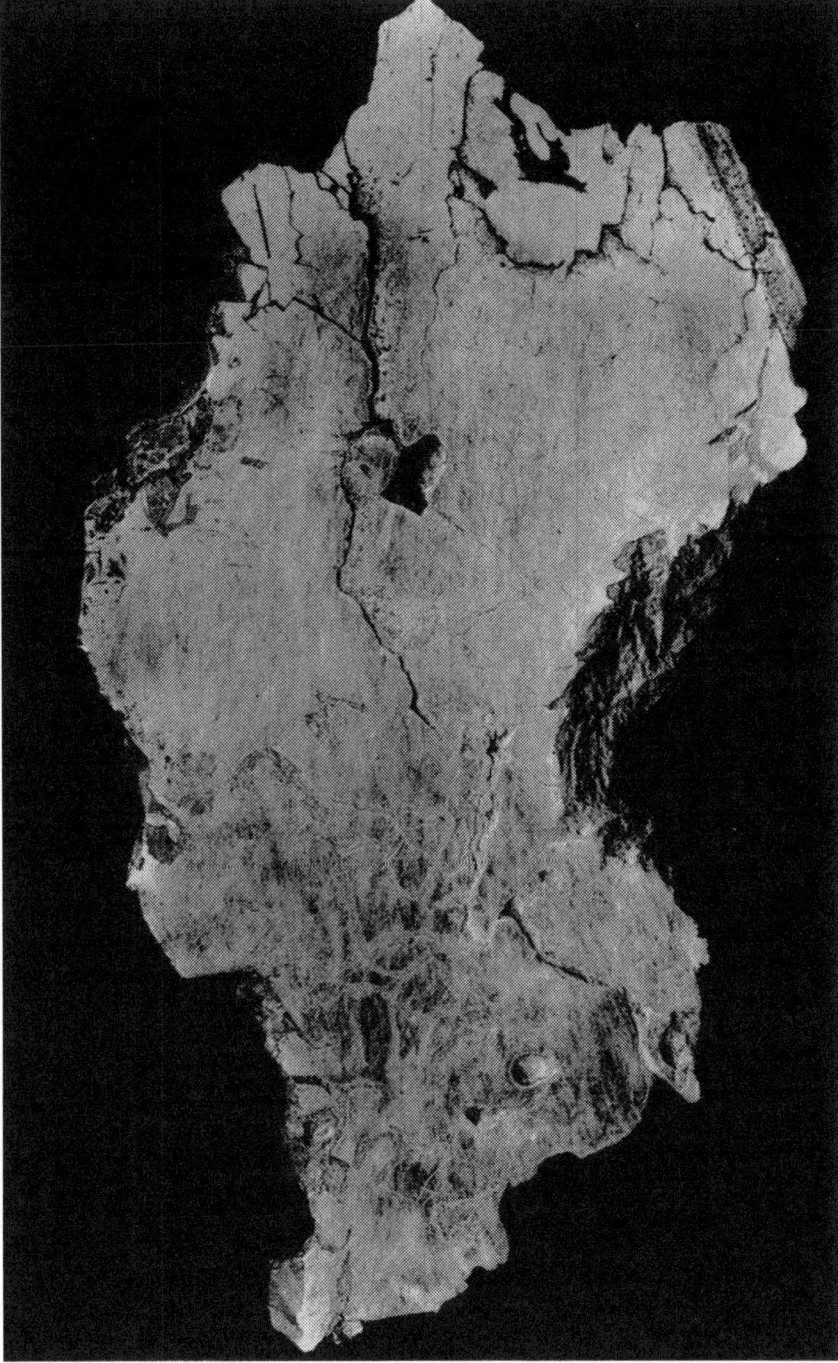

Sao Juliao X $\frac{7}{11}$.

Bacubirito X 1.

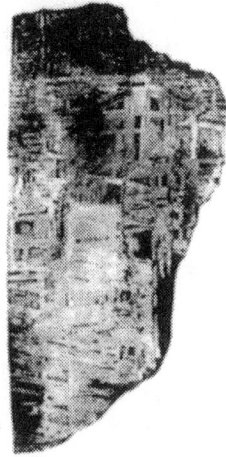

Colfax X 1.

Mukerop X 1

Indian Valley X $\frac{5}{12}$.

Sacramento Mountains X $\frac{5}{12}$

Saint Genevieve County X $\frac{5}{12}$.

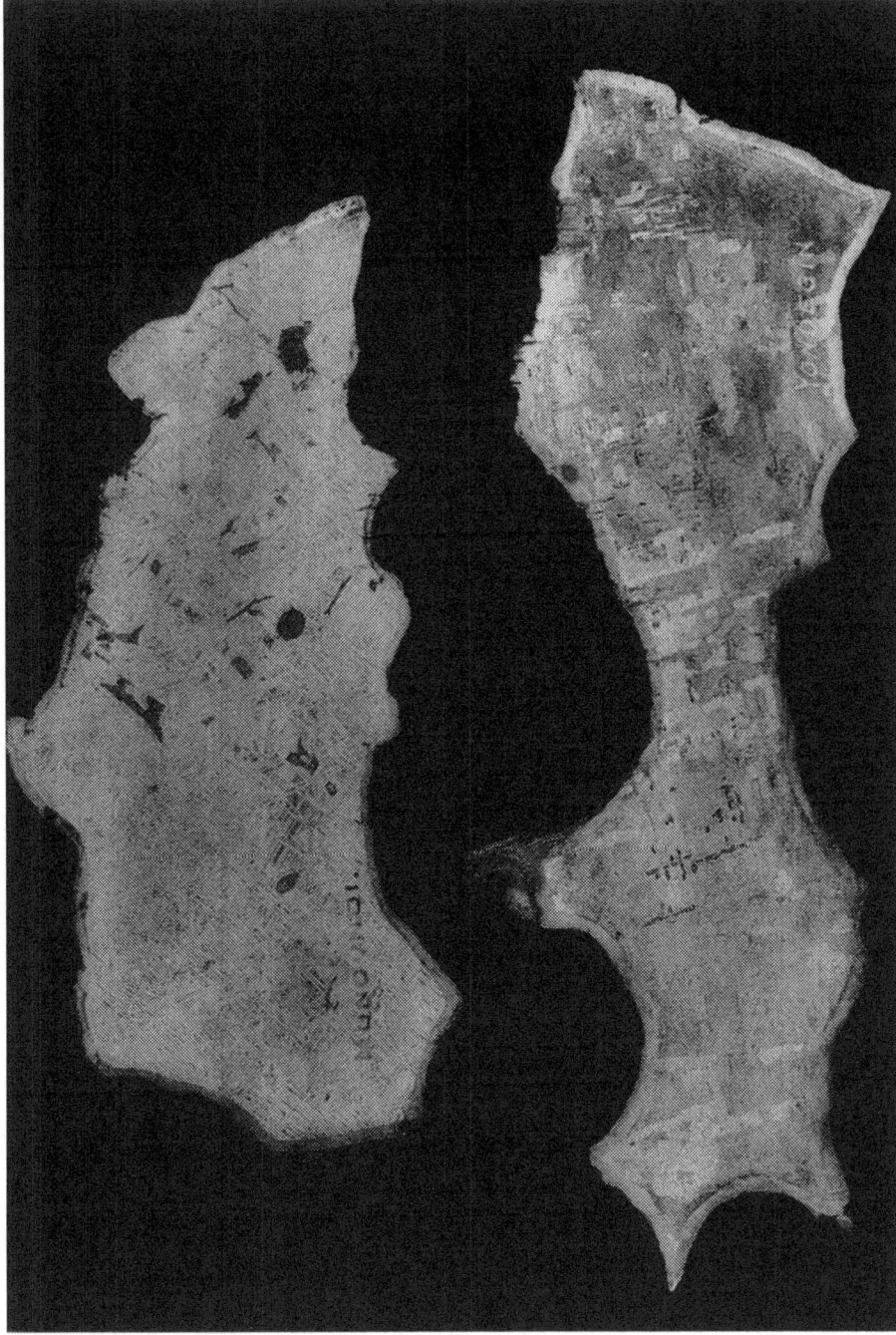

Mungindi × ⅝.

Youndegin × ⅝.

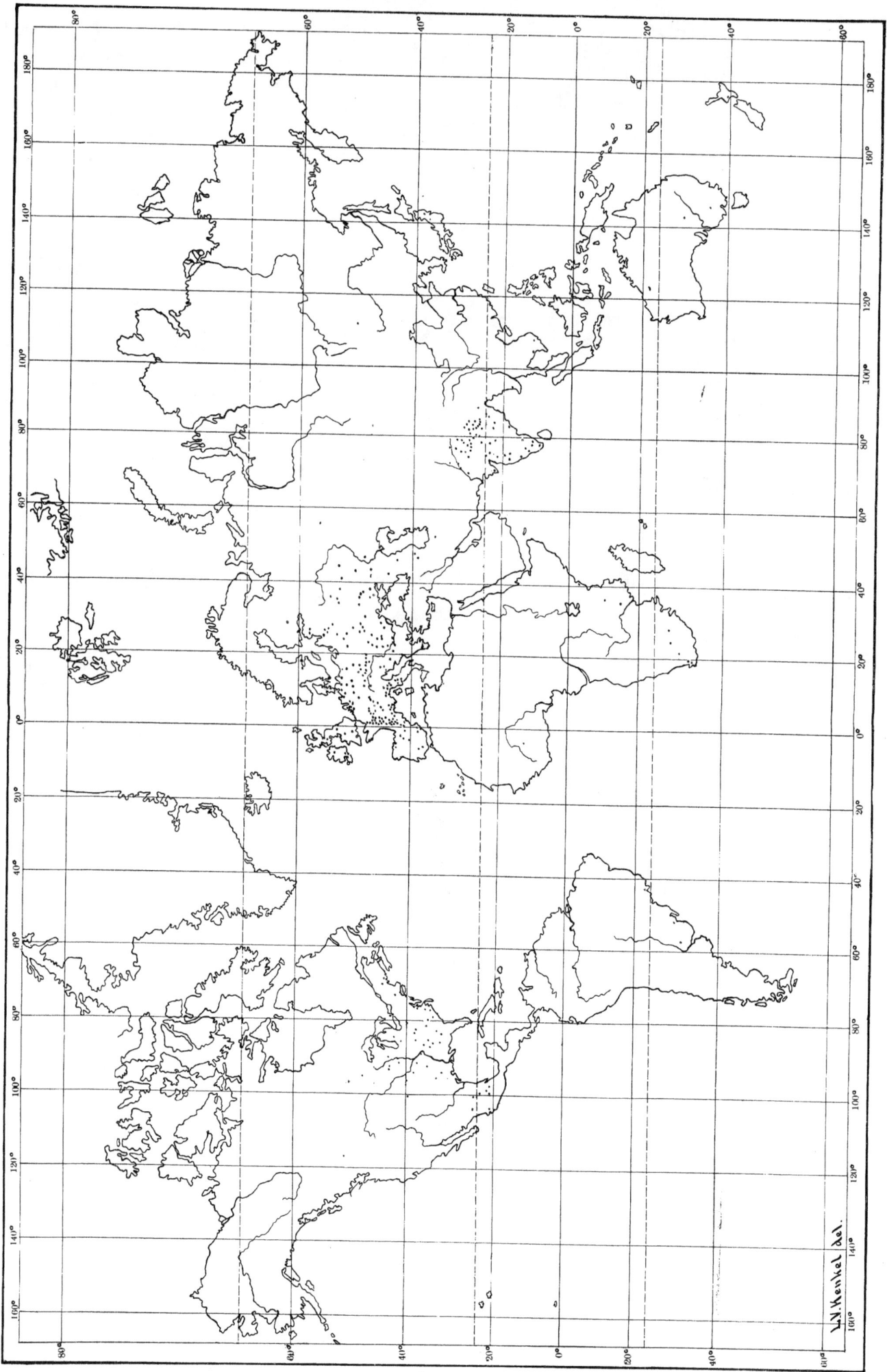

Map of known Meteorite Falls up to 1903.

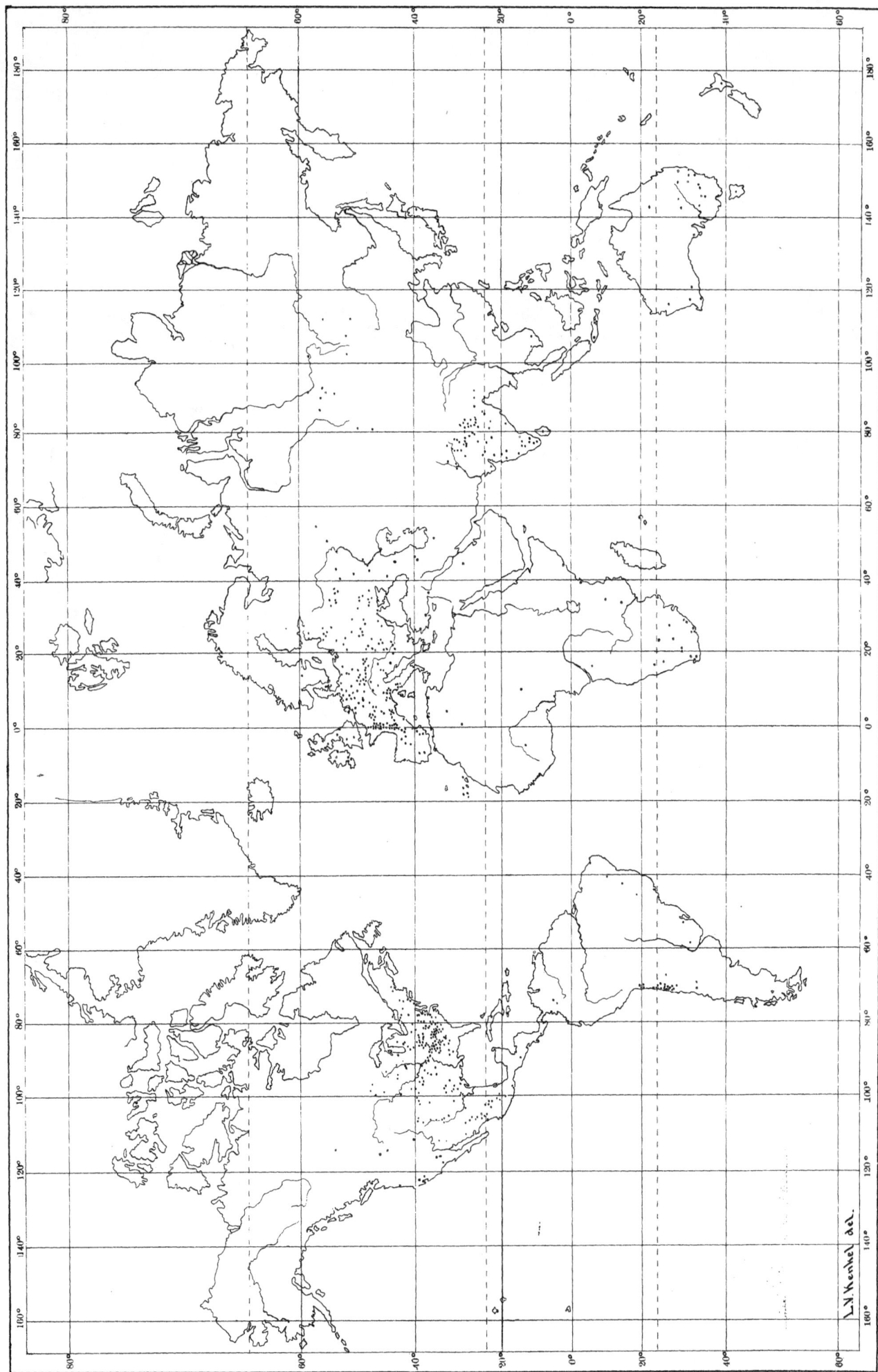

L. W. Menkel del.

Map of known Meteorite Falls and Finds up to 1903.